Compiled by Chinese Academy of Tropical Agricultural Sciences (CATAS) and
Chinese Society for Tropical Crops (CSTC)
A Series of Books for Field Guide to Common Plants in FSM

General Editor: Liu Guodao

Field Guide to Forages in the Federated States of Micronesia

Editors in Chief: Yang Hubiao Zhang Xue

China Agricultural Science and Technology Press

图书在版编目（CIP）数据

密克罗尼西亚联邦饲用植物图鉴 = Field Guide to Forages in the Federated States of Micronesia / 杨虎彪，张雪主编 . —北京：中国农业科学技术出版社，2021.5

（密克罗尼西亚常见植物图鉴系列丛书 / 刘国道主编）

ISBN 978-7-5116-5288-1

Ⅰ.①密… Ⅱ.①杨… ②张… Ⅲ.①牧草—种质资源—密克罗尼西亚联邦—图集 Ⅳ.① S540.24-64

中国版本图书馆 CIP 数据核字（2021）第 068617 号

责任编辑　徐定娜
责任校对　贾海霞
责任印制　姜义伟　王思文

出 版 者	中国农业科学技术出版社 北京市中关村南大街 12 号　邮编：100081
电　　话	（010）82105169（编辑室）（010）82109702（发行部） （010）82109709（读者服务部）
传　　真	（010）82109707
网　　址	http://www.castp.cn
发　　行	各地新华书店
印 刷 者	北京科信印刷有限公司
开　　本	787 mm×1 092 mm　1/16
印　　张	7.5
字　　数	336 千字
版　　次	2021 年 5 月第 1 版　2021 年 5 月第 1 次印刷
定　　价	108.00 元

版权所有·侵权必究

About the Author

Dr. Liu Guodao, born in June 1963 in Tengchong City, Yunnan province, is the incumbent Vice President of Chinese Academy of Tropical Agricultural Sciences (CATAS). Being a professor and PhD tutor, he also serves as the Director-General of the China-Republic of the Congo Agricultural Technology Demonstration Center.

In 2007, he was granted with his PhD degree from the South China University of Tropical Agriculture, majoring in Crop Cultivation and Farming.

Apart from focusing on the work of CATAS, he also acts as a tutor of PhD candidates at Hainan University, Member of the Steering Committee of the FAO Tropical Agriculture Platform (TAP), Council Member of the International Rubber Research and Development Board (IRRDB), Chairman of the Chinese Society for Tropical Crops, Chairman of the Botanical Society of Hainan, Executive Director of the Chinese Grassland Society and Deputy Director of the National Committee for the Examination and Approval of Forage Varieties and the National Committee for the Examination and Approval of Tropical Crop Varieties.

He has long been engaged in the research of tropical forage. He has presided over 30 national, provincial and ministerial-level projects: namely the "National Project on Key Basic Research (973 Program)" and international cooperation projects of the Ministry of Science and Technology, projects of the National Natural Science Foundation of China, projects of the International Center for Tropical Agriculture in Colombia and a bunch of projects sponsored by the Ministry of Agriculture and Rural Affairs (MARA) including the Talent Support Project, the "948" Program and the Infrastructure Project and Special

Scientific Research Projects of Public Welfare Industry.

He has published more than 300 monographs in domestic and international journals such as "New Phytologist" "Journal of Experimental Botany" "The Rangeland Journal" "Acta Prataculturae Sinica" "Acta Agrestia Sinica" "Chinese Journal of Tropical Crops", among which there nearly 20 were being included in the SCI database. Besides, he has compiled over 10 monographs, encompassing "Poaceae Plants in Hainan" "Cyperaceae Plants in Hainan" "Forage Plants in Hainan" "Germplasm Resources of Tropical Crops" "Germplasm Resources of Tropical Forage Plants" "Seeds of Tropical Forage Plants" "Chinese Tropical Forage Plant Resources". As the chief editor, he came out a textbook-*Tropical Forage Cultivation*, and two series of books-*Practical Techniques for Animal Husbandry in South China Agricultural Regions* (19 volumes) and *Practical Techniques for Chinese Tropical Agriculture "Going Global"* (16 volumes).

He has won more than 20 provincial-level and ministerial-level science and technology awards. They are the Team Award, the Popular Science Award and the First Prize of the MARA China Agricultural Science and Technology Award, the Special Prize of Hainan Natural Science Award, the First Prize of the Hainan Science and Technology Progress Award and the First Prize of Hainan Science and Technology Achievement Transformation Award.

He developed 23 new forage varieties including Reyan No. 4 King grass. He was granted with 6 patents of invention and 10 utility models by national patent authorities. He is an Outstanding Contributor in Hainan province and a Special Government Allowance Expert of the State Council.

Below are the awards he has won over the years: in 2020, "the Ho Leung Ho Lee Foundation Award for Science and Technology Innovation"; in 2018, "the High-Level Talent of Hainan province" "the Outstanding Talent of Hainan province" "the Hainan Science and Technology Figure"; in 2015, Team Award of "the China Agricultural Science and Technology Award" by the Ministry of Agriculture; in 2012, "the National Outstanding Agricultural Talents Prize" awarded by the Ministry of Agriculture and as team leader of the team award: "the Ministry of Agriculture Innovation Team" (focusing on the research of Tropical forage germplasm innovation and utilization); in 2010, the first-level candidate of the "515 Talent Project" in Hainan province; in 2005, "the Outstanding Talent of Hainan province"; in 2004, the first group of national-level candidates for the "New Century Talents Project" "the 4th Hainan Youth Science and Technology Award" "the 4th Hainan Youth May 4th Medal" "the 8th China Youth Science and Technology Award" "the Hainan Provincial International Science and Technology Cooperation Contribution Award"; in 2003, "a Cross-Century Outstanding Talent" awarded by the Ministry of Education; In 2001, "the 7th China Youth Science and Technology Award" of Chinese Association of Agricultural Science Societies, "the National Advanced Worker of Agricultural Science and Technology"; in 1993, "the Award for Talents with Outstanding Contributions after Returning to China" by the State Administration of Foreign Experts Affairs.

 Dr. Yang Hubiao, born in September 1983 in Dali City, Yunnan province, is an Associate Professor in the Tropical Crops Genetic Resources Institute of Chinese Academy of Tropical Agricultural Sciences, mainly engaged in plant taxonomy and crop germplasm resources collection. He has collected more than 4,000 grass seedlings, discovered and released 6 new species, recorded 1 new genus and over 40 new species in the world.

He presided over a bunch of projects, such as "Investigation of Grassland and Forage Resources in Southern China" and projects of the National Natural Science Foundation of China; published 22 research papers in "Phytotaxa" "PLoS One" "Acta Botany" and other journals; compiled many research monographs as Chief Editor or Deputy Chief Editor, encompassing *Poaceae Plants in Hainan, Cyperaceae Plants in Hainan, Cultivars of Tropical Forage Plants in Hainan, Forage Plants of Cyperaceae in Hainan.* In 2015, he was included into the Youth Talent Promotion Project of the China Association for science and technology. In 2019, he was awarded the title of State-Advanced Individual of New Variety Protection by the Ministry of Agriculture and Rural Affairs. In 2020, he was selected as the young "Nanhai Expert" of High-Level Talents in Hainan province, and won one special prize of Natural Science Award of Hainan Science and Technology Progress Award in 2020. Now he is the master's graduate tutor of Hainan University, member of Professional Committee for Grassland Resources of China Society of Natural Resources and member of the Sixth Crop Variety Approval Committee of Hainan province.

A Series of Books for Field Guide to Common Plants in FSM

General Editor: Liu Guodao

Field Guide to Forages in the Federated States of Micronesia Editorial Board

Editors in chief:

Yang Hubiao Zhang Xue

Associate editors in chief:

Li Xiaoxia Wang Jinhui Liu Haiqing

Members (in alphabet order of surname):

Fan Haikuo	Gong Shufang	Hao Chaoyun	Huang Guixiu
Li Weiming	Li Xiaoxia	Liu Guodao	Liu Haiqing
Tang Qinghua	Wang Yuanyuan	Wang Qinglong	Wang Jinhui
Wang Xiaofang	Yang Guangsui	Yang Hubiao	You Wen
Zheng Xiaowei	Zhang Xue		

Photographers:

Yang Hubiao Liu Guodao Wang Qinglong

Translator:

Huang Yin

The President
Palikir, Pohnpei
Federated States of Micronesia

FOREWORD

It is with great pleasure that I present this publication, "Agriculture Guideline Booklet" to the people of the Federated States of Micronesia (FSM).

The Agriculture Guideline Booklet is intended to strengthen the FSM Agriculture Sector by providing farmers and families the latest information that can be used by all in our communities to practice sound agricultural practices and to support and strengthen our local, state and national policies in food security. I am confident that the comprehensive notes, tools and data provided in the guideline booklets will be of great value to our economic development sector.

Special Appreciation is extended to the Government of the People's Republic of China, mostly the Chinese Academy of Tropical Agricultural Sciences (CATAS) for assisting the Government of the FSM especially our sisters' island states in publishing books for agricultural production. Your generous assistance in providing the practical farming techniques in agriculture will make the people of the FSM more agriculturally productive.

I would also like to thank our key staff of the National Government, Department of Resources and Development, the states' agriculture and forestry divisions and all relevant partners and stakeholders for their kind assistance and support extended to the team of Scientists and experts from CATAS during their extensive visit and work done in the FSM in 2018.

We look forward to a mutually beneficial partnership.

Sincerely,

David W. Panuelo
President

Preface

Claiming waters of over 3,000 square kilometers, the vast area where Pacific island countries nestle is home to more than 10,000 islands. Its location at the intersection of the east-west and north-south main traffic artery of Pacific wins itself geo-strategic significance. There are rich natural resources such as agricultural and mineral resources, oil and gas here. The relationship between the Federated States of Micronesia (hereinafter referred to as FSM) and China ushered in a new era in 2014 when Xi Jinping, President of China, and the leader of FSM decided to establish a strategic partnership on the basis of mutual respect and common development. Mr Christian, President of FSM, took a successful visit to China in March 2017 during which a consensus had been reached between the leaders that the traditional relationship should be deepened and pragmatic cooperation (especially in agriculture) be strengthened. This visit pointed out the direction for the development of relationship between the two countries. In November 2018, President Xi visited Papua New Guinea and in a collective meeting met 8 leaders of Pacific Island countries (with whom China has established diplomatic relation). China elevated the relationship between the countries into a comprehensive and strategic one on the basis of mutual respect and common development, a sign foreseeing a brand new prospect of cooperation.

The government of China launched a project aimed at assisting FSM in setting up demonstration farms in 1998. Until now, China has completed 10 agricultural technology cooperation projects. To answer the request of the government of FSM, Chinese Academy of Tropical Agricultural Sciences (hereinafter referred to as CATAS), directly affiliated with the

Ministry of Agriculture and Rural Affairs of China, was elected by the government of China to carry out training courses on agricultural technology in FSM during 2017—2018. The fruitful outcome is an output of training 125 agricultural backbone technicians and a series of popular science books which are entitled "Field Guide to Forages in the Federated States of Micronesia" "Field Guide to Flowers and Ornamental Plants in the Federated States of Micronesia" "Field Guide to Medicinal Plants in the Federated States of Micronesia" "Field Guide to Fruits and Vegetables in the Federated States of Micronesia" "Coconut Germplasm Resources in the Federated States of Micronesia" and "Field Guide to Plant Diseases, Insect Pests and Weeds in the Federated States of Micronesia".

In these books, 492 accessions of germplasm resources such as coconut, fruits, vegetables, flowers, forages, medical plants, and pests and weeds are systematically elaborated with profuse inclusion of pictures. They are rare and precious references to the agricultural resources in FSM, as well as a heart-winning project among China's aids to FSM.

Upon the notable moment of China-Pacific Island Countries Agriculture Ministers Meeting, I would like to send my sincere respect and congratulation to the experts of CATAS and friends from FSM who have contributed remarkably to the production of these books. I am firmly convinced that the exchange between the two countries on agriculture, culture and education will be much closer under the background of the publication of these books and Nadi Declaration of China-Pacific Island Countries Agriculture Ministers Meeting, and that more fruitful results will come about. I also believe that the team of experts in tropical agriculture mainly from the CATAS will make a greater contribution to closer connection in agricultural development strategies and plans between China and FSM, and closer cooperation in exchanges and capacity-building of agriculture staffs, in agricultural science and technology for the development of agriculture of both countries, in agricultural investment and trade, in facilitating FSM to expand industry chain and value chain of agriculture, etc.

Qu Dongyu
Director General
Food and Agriculture Organization of the United Nations
July 23, 2019

Located in the northern and central Pacific region, the Federated States of Micronesia (FSM) is an important hub connecting Asia and America. Micronesia has a large sea area, rich marine resources, good ecological environment, and unique traditional culture.

In the past 30 years since the establishment of diplomatic relations between China and FSM, cooperation in diverse fields at various levels has been further developed. Since the 18th National Congress of the Communist Party of China, under the guidance of Xi Jinping's thoughts on diplomacy, China has adhered to the fine diplomatic tradition of treating all countries as equals, adhered to the principle of upholding justice while pursuing shared interests and the principle of sincerity, real results, affinity, and good faith, and made historic achievements in the development of P.R. China-FSM relations.

The Chinese government attaches great importance to P.R. China-FSM relations and always sees FSM as a good friend and a good partner in the Pacific island region. In 2014, President Xi Jinping and the leader of the FSM made the decision to build a strategic partnership featuring mutual respect and common development, opening a new chapter of P.R. China-FSM relations. In 2017, FSM President Peter Christian made a successful visit to China. President Xi Jinping and President Christian reached broad consensuses on deepening the traditional friendship between the two countries and expanding practical cooperation between the two sides, and thus further promoted P.R. China-FSM relations. In 2018, Chinese President Xi Jinping and Micronesian President Peter Christian had a successful meeting again in PNG and made significant achievements, deciding to upgrade P.R. China-FSM relations to a new stage of Comprehensive Strategic Partnership, thus charting the course for future long-term development of P.R. China-FSM relations.

In 1998, the Chinese government implemented the P.R. China-FSM demonstration farm project in FSM. Ten agricultural technology cooperation projects have been completed, which has become the "golden signboard" for China's aid to FSM. From 2017 to 2018, the Chinese Academy of Tropical Agricultural Sciences (CATAS), directly affiliated with the Ministry of Agriculture and Rural Affairs, conducted a month-long technical training on pest control of coconut trees in FSM at the request of the Government of FSM. 125 agricultural managers, technical personnel and growers were trained in Yap, Chuuk, Kosrae and Pohnpei, and the biological control technology demonstration of the major dangerous pest, Coconut Leaf Beetle, was carried out. At the same time, the experts took advantage of the spare time of the training course and spared no effort to carry out the preliminary evaluation of the investigation and utilization of agricultural resources, such as coconut, areca nut, fruit tree, flower, forage, medicinal plant, melon and vegetable, crop disease, insect pest and weed diseases, in the field in conjunction with Department of Resources and Development of FSM and the vast number of trainees, organized and compiled a series of popular science books, such as "Field Guide to Forages in the Federated States of Micronesia" "Field Guide to Flowers and Ornamental Plants in the Federated States of Micronesia" "Field Guide to Medicinal Plants in the Federated States of Micronesia" "Field Guide to Fruits and Vegetables in the Federated States of Micronesia" "Coconut Germplasm Resources in the Federated States of Micronesia" and "Field Guide to Plant Diseases, Insect Pests and Weeds in the Federated States of Micronesia".

The book introduces 37 kinds of coconut germplasm resources, 60 kinds of fruits and vegetables, 91 kinds of angiosperm flowers as well as 13 kinds of ornamental pteridophytes, 100 kinds of forage plants, 117 kinds of medicinal plants, 74 kinds of crop diseases, pests and weed diseases, in an easy-to-understand manner. It is a rare agricultural resource illustration in FSM. This series of books is not only suitable for the scientific and educational workers of FSM, but also it is a valuable reference book for industry managers, students, growers and all other people who are interested in the agricultural resources of FSM.

This series is of great significance for it is published on the occasion of the 30th anniversary of the establishment of diplomatic relations between the People's Republic of

China and FSM. Here, I would like to pay tribute to the experts from CATAS and the friends in FSM who have made outstanding contributions to this series of books. I congratulate and thank all the participants in this series for their hard and excellent work. I firmly believe that based on this series of books, the agricultural and cultural exchanges between China and FSM will get closer with each passing day, and better results will be achieved more quickly. At the same time, I firmly believe that the Chinese Tropical Agricultural Research Team, with CATAS as its main force, will bring new vigour and make new contributions to promoting the in-depth development of the strategic partnership between the People's Republic of China and the Federated States of Micronesia, strengthening solidarity and cooperation between P.R. China and the developing countries, and the P.R. China-FSM joint pursuit of the Belt and Road initiative and building a community with a shared future for the humanity.

Ambassador Extraordinary & Plenipotentiary of
the People's Republic of China to
the Federated States of Micronesia
May 23, 2019

Foreword

The Federated States of Micronesia (hereinafter referred to as the Micronesia) is a resource-rich, culturally unique federal island nation located in the Western Pacific Ocean, consisting of Yapese, Chuuk, Kosrae, and the capital, Pohnpei. Since the establishment of diplomatic relations between China and Micronesia, the two countries have been deepening their traditional friendship, cooperation and exchanges. In July 2018, according to the cooperation needs of agricultural technology in Micronesia, Chinese Academy of Tropical Agricultural Sciences, empowered by the Chinese government, sent a research group to Micronesia to carry out a "Belt and Road" technical assistance service for tropical agriculture. It is very fortunate that the authors had a chance to visit this country to have a good picture of local natural ecology and agricultural development.

The climate of Micronesia belongs to tropical oceanic climate, with high vegetation coverage, mainly coastal forests. It is of a huge potential for development and utilization of abundant plant species diversity and specific tropical crop germplasm resources that are highly closely related to geographical conditions and climatic characteristics. However, due to its unique history and culture, excellent germplasm resources have been in a wild state. With the development of today's agriculture, Micronesia has put forward urgent technical needs.

Sharing experience and technology is an important mission of the "Belt and Road" initiative. With this in mind, combined with the development requirements of the grassland livestock industry of Micronesia, the authors completed the survey and research on the development level of forage resources and animal husbandry in the country, and found that there were abundant forage resources in the four states, but the development of animal husbandry was lagging behind. The demand for livestock, poultry and dairy products had long been dependent on imports. However, in fact, Micronesia had an excellent advantage in the development of animal husbandry, which was characterized by the availability of abundant grassland resources, and a large number of excellent forages such as the Paddle grass (*Ischaemum polystachyum* J. Presl)s, Angel grass (*Paspalum paniculatum* L) and beach pea (*Vigna marina* (Bum.) Merr). Liu Guodao, Vice President of the Chinese Academy of Tropical Agricultural Sciences, said to President Peter Christian, "The grass we see in Micronesia is milk and meat in the future", which resonated with the senior officials of Micronesia, and his proposal to develop animal husbandry was highly valued by President Peter Christian.

Firmly adhering to the important mission of the "Belt and Road" to meet the development needs of Micronesia and actively promote the future development of its animal husbandry, the authors systematically sorted out the resources of the forage plants in Micronesia for later use. This survey was funded by the Belt and Road Tropical project and the project of the Young Talents of the China Association for Science and Technology. This survey was very successful. Thanks are due to the Chinese Embassy in Micronesia, the accompanying officials of Micronesia and the research group of the Chinese Academy of Tropical Agricultural Sciences for their helpful support.

General Editor

Vice President of Chinese Academy of Tropical Agricultural Sciences

March 22, 2019

Contents

Acacia auriculiformis 1	*Desmodium tortuosum* 23
Acacia confusa 2	*Desmodium heterophyllum* 24
Calliandra haematocephala 3	*Stylosanthes hamata* 25
Calliandra surinamensis 4	*Aeschynomene indica* 26
Pithecellobium dulce 5	*Aeschynomene americana* 27
Leucaena leucocephala 6	*Arachis hypogaea* 28
Desmanthus virgatus 7	*Flemingia macrophylla* 29
Bauhinia purpurea 8	*Flemingia strobilifera* 30
Cassia fistula 9	*Crotalaria pallida* 31
Caesalpinia bonduc 10	*Crotalaria incana* 32
Senna tora 11	*Crotalaria assamica* 33
Senna alata 12	*Kummerowia striata* 34
Senna occidentalis 13	*Cajanus cajan* 35
Senna surattensis 14	*Macroptilium atropurpureum* 36
Chamaecrista mimosoides 15	*Lablab purpureus* 37
Abrus precatorius 16	*Psophocarpus tetragonolobus* 38
Alysicarpus vaginalis 17	*Vigna unguiculata* 39
Dendrolobium umbellatum 18	*Vigna marina* 40
Tadehagi triquetrum 19	*Clitoria ternatea* 41
Desmodium triflorum 20	*Canavalia cathartica* 42
Desmodium heterocarpon 21	*Calopogonium mucunoides* 43
Desmodium incanum 22	*Sesbania grandiflora* 44

Sesbania cannabina	45	*Coix lacryma-jobi*	73
Sporobolus fertilis	46	*Miscanthus floridulus*	74
Eleusine indica	47	*Ischaemum polystachyum*	75
Arundo donax	48	*Ischaemum ciliare*	76
Eragrostis atrovirens	49	*Bothriochloa ischaemum*	77
Eragrostis unioloides	50	*Cymbopogon citratus*	78
Centotheca lappacea	51	*Schoenus calostachyus*	79
Zoysia japonica	52	*Rhynchospora rubra*	80
Zoysia matrella	53	*Fimbristylis tristachya*	81
Chloris formosana	54	*Fimbristylis cymosa*	82
Cynodon dactylon	55	*Bidens pilosa*	83
Thuarea involuta	56	*Crassocephalum crepidioides*	84
Setaria geniculata	57	*Tridax procumbens*	85
Pennisetum purpureum	58	*Wollastonia biflora*	86
Pennisetum polystachion	59	*Elephantopus tomentosus*	87
Cenchrus echinatus	60	*Emilia sonchifolia*	88
Echinochloa colona	61	*Ageratum conyzoides*	89
Oplismenus compositus	62	*Pilea microphylla*	90
Digitaria microbachne	63	*Pouzolzia zeylanica*	91
Axonopus compressus	64	*Ipomoea pes-caprae*	92
Paspalum scrobiculatum	65	*Vitex trifolia* var. *simplicifolia*	93
Paspalum paniculatum	66	*Ipomoea aquatica*	94
Paspalum vaginatum	67	*Ipomoea batatas*	95
Paspalum dilatatum	68	*Asystasia gangetica*	96
Eriochloa procera	69	*Colocasia esculenta*	97
Brachiaria mutica	70	*Cyrtosperma merkusii*	98
Panicum maximum	71	*Artocarpus communis*	99
Panicum repens	72	*Trema cannabina*	100

Acacia auriculiformis

Pohnpei: Tuhkehn pwelmwahu

Trees, evergreen, upto 20 m tall. Bark gray-white, smooth. Branches pendulous; branchlets angular, glabrous, with conspicuous lenticels. Phyllodes falcate-oblong, (10–20) cm × [1.5–4(–6)] cm, conspicuous main veins 3 or 4, both ends attenuate. Spikes 1 to several, fasciculate, axillary or terminal, 3.5–8 cm. Flowers orange-yellow. Calyx 0.5–1 mm, shallowly dentate at apex. Petals oblong, 1.5–2 mm long. Filaments 2.5–4 mm long. Ovary densely puberulent. Legume contorted when mature, (5–8) cm × (0.8–1.2) cm, valves woody. Seeds ca. 12, black, ca. 5 × 3.5 mm.

Plant parts

Distribution: Pohnpei.

Utilization: Beautiful colors, mostly planted as a street tree. It is also a fast-growing tree species for pulp. Sheep also like to eat its young leaves.

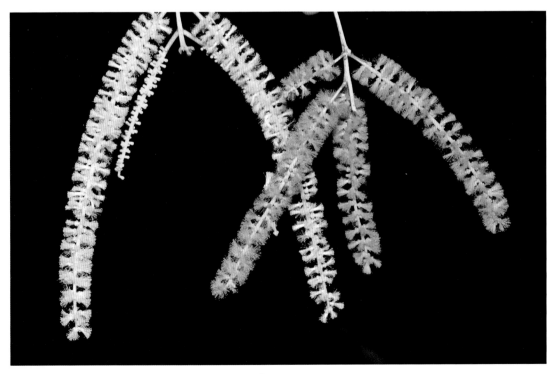
Inflorescence

Acacia confusa

Pohnpei: Pilampwoia

English: Formosa acacia, Formosan koa

Trees, evergreen, 6–15 m tall, glabrous. Branches gray or brown, unarmed; branchlets slender, angular. Phyllodes straight or slightly falcate, linear-lanceolate, (6–10) cm × (0.5–1.3) cm, leathery, both surfaces glabrous, conspicuous longitudinal veins (3–)5(–8), both ends acuminate, apex slightly obtuse, usually with hooked acumen. Heads solitary or 2- or 3- fasiculate, axillary, globose, ca. 0.7 cm in diam.; peduncles 7–13 mm, slender.

Plant

Flowers golden yellow, fragrant. Calyx 1–1.3 mm. Petals greenish, 1.5–1.9 mm. Stamens numerous, ca. 3.5 mm. Ovary yellow-brown villous; style ca. 4 mm. Legume black-brown, flat, [4–9(–12)] cm × (0.7–1) cm. Seeds broadly elliptic, (5–6) mm × (3.5–5) mm.

Distribution: Yapese, Pohnpei.

Utilization: It is mostly used for planting as a street tree. It is also a fast-growing tree species for pulp. Sheep also like to eat its young leaves.

Inflorescence

Calliandra haematocephala

English: Blood red tassel flower

Shrubs or small trees, deciduous, 1–3 m tall. Branchlets brown, cylindric, rough. Stipules persistent, ovate-lanceolate; petiole 1–2.5 cm; pinnae 1 pair, 8–13 cm long; petiolules ca. 1 mm; leaflets 7–9 pairs, obliquely lanceolate, (2–4) cm × (7–15) mm, sparsely pilose along margin, midvein close to upper margin, base oblique, apex obtuse, mucronate. Heads axillary, ca. 3 cm in diam. (including filaments); peduncles 1–3.5 cm. Calyx campanulate, ca. 2 mm long. Corolla purplish red; tube 3.5–5 mm long, 5-lobed; lobes reflexed, ca. 3 mm long, glabrous. Stamens numerous, scarlet, very brilliant; staminal tube white, ca. 6 cm long, mouth inside with a subulate appendage; filaments dark red, ca. 2 cm long. Legume dull brown, linear-oblanceolate, (6–11) cm × (5–13) mm, valves elastically open from apex to base along sutures when ripe, reflexed. Seeds 5 or 6, brown, oblong, (7–10) × ca. 4 mm.

Plant

Distribution: Pohnpei, Yapese, Chuuk, Kosrae

Utilization: It is large in leaf volume, good in palatability and high in protein content, and can be used as feed.

Inflorescence

Calliandra surinamensis

English: Pink powder puff, Surinam powderpuff

Shrub or small tree about 2 m high (up to 6 m elsewhere). Branches spreading; branchlets cylindrical, brown, rough. Leaves bipinnately compound, not glandular; pinnae 1 (infrequently 2 or 3) pair; leaflets alternate. Flowers are sessile in showy heads, axillary or terminal; calyx campanulate, shallowly lobed; petals confluent to the middle; central flower usually abnormal, with a long tubular corolla; stamen numerous, red or white, long protrudent, confluent to tubes at lower part; anther usually with glandular hair; carpel 1, sessile; ovules numerous; stigma linear. Pods linear, flattened, straight or slightly curved, usually narrow at base, thick margined; carpal dehisces into two from the top to the basal sulture. Seeds obovate or oblong, compressed; testa hard, with pleurogram; aril absent.

Plant

Distribution: Pohnpei, Yapese, Chuuk, Kosrae

Utilization: It is a horticultural ornamental plant and is mostly used for landscaping. It can also be used as a forage since it is a legume plant with a high protein content.

Bud

Inflorescence

Pithecellobium dulce

Chuuk: Kamachuri

Pohnpei: Gamachil

Trees, evergreen. Branches often pendulous; branchlets armed with spinescent stipules. Pinnae 1 pair; glands at junction of pinnae and leaflets; leaflets sessile, 1 pair per pinna, elliptic or obovate-elliptic, (2–5) cm × (0.2–2.5) cm, both surfaces glabrous, variable in size, reticulate veins raised abaxially, base slightly oblique, apex obtuse or emarginate. Inflorescence pedunculate heads, aggregated in terminal panicles. Calyx funnel-shaped, 1–1.5 mm, tomentose. Corolla white or pale yellow, ca. 6 mm, densely tomentose. Stamens numerous, 8–10 mm long, connate

Plant parts

into a tube at base. Pods dark red, curved into a circle, flat, 10–13 cm long, about 1 cm in diam, dilated. Seeds dark brown, shiny, ovoid-ellipsoid, ca. 1.5 cm, hard, with pleurogram.

Distribution: Pohnpei, Yapese, Chuuk, Kosrae

Utilization: It has many uses. Its wood is a good building material, its leaves and pods can be used as feed, while the aril is sweet and sour and can be used to make beverages.

Inflorescence

Leucaena leucocephala

Pohnpei: Dangandangan, Tangantangan

Yapese: Ganitnityuwan, Tangantan

Shrubs or small trees, 2–6 m tall. Branchlets pubescent, glabrous when old, with brown lenticels. Stipules caducous, deltoid, very small; pinnae 4–8 pairs, 5–9(–16) cm, rachis pubescent with black glands at location of lowest pinnae; leaflets 5–15 pairs, linear–oblong, (7–13) mm × (1.5–3) mm, main vein close to upper margin, base cuneate, margin ciliate, apex acute. Heads usually 1 or 2, axillary, 2–3 cm in diam.; peduncle 2–4 cm; bracts deciduous, pubescent. Flowers white. Calyx ca. 3 mm, abaxially glabrous at base, puberulent at apex, 5- toothed. Petals narrowly oblanceolate, ca. 5 mm, abaxially pubescent. Stamens 10, sparsely pubescent, ca.

Seeds

7 mm. Ovary shortly stipitate, sparsely pubescent at upper part; stigma cupular. Legume straight, strap-shaped, flat, (10–18) cm × (1.4–2) cm, leathery, base attenuate, pedicel to 3 cm, puberulent, beak acute, hard. Seeds 6–25, brown, glossy, narrowly ovoid, flat, (6–9) mm × (3–4.5) mm.

Distribution: Pohnpei, Yapese, Chuuk, Kosrae

Utilization: Tender stems and leaves have good palatability and are rich in protein, carotene and vitamins. They can be used as feed for cattle and sheep. Leaf meal is an excellent supplementary feed for pigs, rabbits and poultry.

Plant

Fruit Pods

Desmanthus virgatus

Latin: *Desmanthus virgatus* (L.) Willd.

English: Slender mimosa, Virgate mimosa, Wild tan-tan

Perennial shrubby herb, 0.5–1.3 m tall; branches slender, ribbed, pubescent on ribs. Stipules setose, 3–6 mm long. Leaves bipinnately compound; pinnae 2–4 pairs of feathers, 1.2–2.5 cm long; leaflets 6–21 pairs, oblong, 4–6 mm long, about 2 mm wide. Inflorescence head, 5 mm in diameter, green-white, 4–10 flowered; peduncle 1–4 cm long; bracteoles ovate, long pointed; calyx bell-shaped, about 2 mm long, with short calyx teeth; narrowly oblong petals narrowly oblong, about 3 mm long; stamens 10. Pods linear, 4–11 cm long, 2–4 mm wide; seeds oblique, 2.5–3 mm long.

Plant

Distribution: Pohnpei, Chuuk, Kosrae

Utilization: Tender stems and leaves have better palatability and higher crude protein content, which are suitable to use as feed for cattle and sheep.

Inflorescence

Bauhinia purpurea

English: Butterfly tree, Butterfly-orchid-tree

Trees or erect shrubs, 7–10 m tall. Bark grayish to dark brownish, thick, smooth; branches puberulent when young, later glabrous. Petiole 3–4 cm; leaf blade suborbicular, (10–15) cm × (9–14) cm, stiffly papery, abaxially almost glabrous, adaxially glabrous, primary veins 9–11, secondary and higher order veins protruding, base shallowly cordate, apex bifid to 1/3–1/2, lobes slightly acute or rarely rounded at apex. Inflorescence a raceme with few flowers, or a panicle with up to 20 flowers, axillary or terminal. Flower buds fusiform, 4-or 5-ridged, with an obtuse apex. Pedicel 7–12 mm. Calyx open as a spathe into 2 lobes, one with 2 teeth and other 3-toothed. Petals light pink, oblanceolate,

Seed

4–5 cm, clawed. Fertile stamens 3; filaments ca. as long as petals. Staminodes 5 or 6, 6–10 mm. Ovary stalked, velvety; style curved; stigma slightly enlarged, obliquely peltate. Legume linear, slightly falcate, flat, (12–25) cm × (2–2.5) cm, dehisced when mature; valves woody. Seeds compressed, sub-orbicular, 12–15 mm in diam; testa dark brown.

Distribution: Pohnpei

Utilization: Tender leaves are of good palatability, contain high crude protein content and can be used as feed for cattle and sheep. In dry season, its dead leaves can also be used as hay.

Plant parts

Cassia fistula

Trees, deciduous, to 15 m tall. Branches slender; barks glabrous and gray when young, rough and dark brown when old. Leaves 30–40 cm, with 3 or 4 pairs of leaflets; leaflets adaxially shiny, broadly ovate or ovate-oblong, (8–13) cm × (4–8) cm, thin leathery, both surfaces puberulent when young, glabrous when mature, base broadly cuneate, apex acute. Racemes axillary, 20–40(–60) cm, lax, pendent, many flowered; flowers 3.5–4 cm in diam. Pedicels 3–5 cm, slender. Sepals narrowly ovate, 1–1.5 cm, reflexed at anthesis. Petals golden yellow, broadly ovate, subequal, 2.5–3.5 cm, shortly clawed. Stamens 10, 3 long with curved filaments 3–4 cm, anthers ca. 5 mm, exceeding petals, 4 short with straight filaments 6–10 mm, reduced stamens with minute anthers. Ovary stalked, strigulose; stigma small. Legume pendulous, blackish brown, terete, sausage-shaped, indehiscent, 30–60 cm, 2–2.5 cm in diam. Seeds numerous, separated by papery septa, glossy brown, elliptic, flattened.

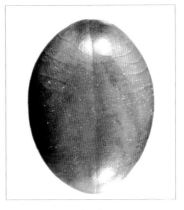

Seed

Distribution: Pohnpei

Utilization: It contains high crude protein, and can be used as feed for cattle and sheep. In dry season, its dead leaves can also be used as hay.

Plant parts

Caesalpinia bonduc

Chuuk: Nickaeoo

Pohnpei: Kehsaphl

English: Beach nicker, Bonduc, Divi-divi

Plant

Climbers, prickly, yellowish pubescent throughout. Prickles straight or somewhat recurved. Leaves 30–45 cm; rachis with recurved prickles; pinnae 6–9 pairs, opposite; stipules deciduous, large, leaflike, usually lobed, lobes to 2 cm; leaflets 6–12 pairs, oblong, (1.5–4) cm × (1.2–2) cm, membranous, both surfaces pubescent, base oblique, apex rounded to acute, mucronate. Racemes axillary, long pedunculate, densely flowered in upper part and sparsely flowered in lower part; bracts caducous at anthesis, reflexed, subulate, 6–8 mm, pubescent. Pedicels 3–5 mm. Sepals 5, ca. 8 mm, both surfaces ferruginous hairy. Petals yellowish; standard tinged with red spots, oblanceolate, clawed. Filaments short, hairy in basal part. Ovary hairy. Legume oblong, (5–7) cm × (4–5) cm, leathery, apex rounded with beak, swollen, with dense, slender spines 5–10 mm. Seeds 2 or 3, grayish, shiny, subglobose.

Distribution: Pohnpei, Chuuk, Yapese, Kosrae

Utilization: Tender leaves are occasionally eaten by sheep.

Fruit pods

Senna tora

English: Foetid cassia, Java bean

Herbs, suffrutescent, annual, erect, 1–2 m tall. Leaves 4–8 cm long; stipules caducous, linear, 10–15 mm, puberscent; petiole without glands; rachis with a club-shaped gland between leaflets; petiolules 1.5–2 mm; leaflets 3 pairs, obovate or obovate-oblong, (2–6) × (1.5–2.5) cm, membranous, abaxially pubescent, adaxially sparsely pubescent, base cuneate to rounded and oblique, apex rounded, cuspidate. Racemes axillary, short, 1- or 2(or 3)-flowered; peduncles 6–10 mm; bracts linear, acute. Pedicels 1–1.5 cm. Sepals ovate or ovate-oblong, 5–8 mm, membranous, outside pubescent. Petals yellow, unequal, obovate, lower 2 slightly longer, (12–15) mm × (5–7) mm, shortly clawed. Fertile stamens 7, subequal, 3 larger, 4 smaller; filaments 1.5–2 mm; anthers square, opening by apical pores, ca. 4 mm; staminodes absent. Ovary sessile, densely white pubescent; style glabrous. Legume terete, subtetragonous, slender, (10–15) cm × (0.3–0.5) cm, both ends acuminate, valves membranous. Seeds 20–30, glossy, rhomboid, ca. 5 mm × 3 mm, with an areole.

Seed

Distribution: Pohnpei

Utilization: It is a good green manure plant and can also be used as feed.

Plant

Flower

Senna alata

Chuuk: Arakak, Arekak, Salai
Pohnpei: Tuhken kilinwai
Yapese: Geking sepan

Shrubs, 1.5–3 m tall. Branches greenish, thick, pubescent. Leaves 30–60 cm; stipules persistent, triangular, 6–10(–15) mm; petiole and rachis with 2 longitudinal ribs and narrow wings; petiolar glands absent; petiolules very short or leaflets subsessile; leaflets 6–12(–20) pairs, oblong or obovate-oblong, (6–15) cm × (3.5–7.5) cm, thinly leathery, glabrous, base obliquely truncate, apex obtusely rounded and cuspidate. Racemes axillary, dense, many flowered, or sometimes several racemes forming a terminal panicle, 10–50 cm, long pedunculate, solitary or branched; peduncles 7–14 cm; bracts caducous, strobilaceous, oblong to broadly ovate, (2–3) cm × (1–2) cm, at first enveloping flowers. Flowers ca. 2.5 cm in diam. Sepals orange-yellow, oblong, unequal. Petals bright yellow, tinged with conspicuous purple veins, ovate–orbicular, (16–24) mm × (10–15) mm, shortly clawed. Stamens 10, fertile stamens 7, opening with apical pores, lower 2 with stout filaments ca. 4 mm and larger anthers, 4 with filaments ca. 2 mm and smaller anthers, reduced stamens 3 or 4. Ovary puberulent, sessile; ovules many. Legume winged, sharply tetragonal, (10–20) cm × (1.5–2) cm, glabrous, with a broad, membranous wing down middle of each valve; wings 4–8 mm wide, papery, crenulate. Seeds 50–60, compressed, deltoid.

Distribution: Pohnpei, Yapese, Chuuk, Kosrae

Utilization: It has large biomass, good palatability and high protein content, and can be used as feed.

Leaves

Inflorescence

Seeds

Plants

Senna occidentalis

Chuuk: Afanafan

English: Antbush, Arsenic bean

Subshrubs or shrubs, erect, 0.8–1.5 m tall, glabrous, few branched. Branches herbaceous, ribbed; roots blackish. Leaves ca. 20 cm; stipules caducous, triangular to lanceolate, 1–2 cm, membranous; petiole 3–4 cm, with a large, brown, ovoid gland near base; petiolule 1–2 mm, with a rotten smell when kneaded; leaflets 3–5(or 6) pairs, ovate to ovate–oblong, (4–10) cm × (2–3.5) cm, membranous, base rounded, apex acuminate. Corymbose racemes few flowered, axillary or terminal, ca. 5 cm; bracts caducous, linear-lanceolate. Flowers ca. 2 cm. Sepals unequal, outer ones suborbicular, ca. 6 mm in diam., inner ones ovate, 8–9 mm. Petals yellow, purplish veined, 2 outer slightly larger, shortly clawed. Fertile stamens 7, anthers opening by apical pores, reduced stamens 3, without anthers or with tiny anthers. Ovary tomentose; style glabrous. Legume brown, with pale thick margins, strap-shaped, falcate, flattened, (10–13) cm × ca. 1 cm, with septa between seeds. Seeds 30–40, flat, orbicular, 3–4 mm in diam.

Distribution: Pohnpei, Chuuk

Utilization: It contains high protein and is prefered by cattle and sheep and other livestock. It is a good plant for fodder.

Plant　　　　　　　　　　　Flower

Inflorescence　　　　　　　　Seeds

Senna surattensis

English: Glaucous cassia, Scrambled eggs bush

Shrubs or small trees, 5–7 m tall. Bark grayish brown, smooth; young branches, petioles, and rachises of leaves puberulent. Leaves 10–15 cm, with 2 or 3 clavate, long glands 1–2 mm on rachis between lowest 2 or 3 pairs of leaflets and in upper part of petiole; stipules subpersistent, linear, 5–10 mm; leaflets 6–9 pairs, abaxially farina-white, ovate to ovate-oblong, (2–5) cm × (1–1.7) cm, abaxially sparsely pubescent, adaxially glabrous, base rounded, apex rounded, slightly emarginate. Racemes in axils of apical leaves, 3–6 cm, 10-15-flowered; peduncles 2.5–5 cm; bracts ovate-oblong, 5–8 mm, outside puberulent, finally reflexed. Pedicels 1–2 cm. Sepals unequal, 2 outer suborbicular, ca. 3 mm in diam., 3 inner obovate, to 7 mm. Petals bright yellow to deep yellow, subequal, ovate to obovate, 1.5–2 cm, with long claw 1–1.5 mm. Stamens 10, all fertile, with short, thick filaments, lowest 2 filaments longer; anthers oblong, subequal, 5–7 mm, opening by apical slits. Ovary puberulent; style glabrous. Legume flat, strap-shaped, dehiscent, (7–10) cm × (0.8–1.5) cm, long slender beaked, valves papery. Seeds 10–25, glossy, flattened.

Seed

Distribution: Kosrae

Utilization: It contains high protein content, preferred by cattle and sheep and other livestock, and is hence a good plant for forage.

Plant

Chamaecrista mimosoides

English: Five-leaf cassia, Japanese tea

Herbs, suffrutescent, annual or perennial, with woody base, 30–60 cm tall, or low shrubs to 1 m tall. Branches, numerous, slender, puberulent. Leaves 4–8 cm, with an orbicular, discoid, sessile gland in upper part of petiole, below lowest pair of leaflets; stipules persistent, linear-subulate, 4–7 mm, with conspicuous longitudinal veins; rachis not canaliculate, sparsely pubescent; leaflets sessile, 20–50(–80) pairs, reddish brown when dry, linear-falcate, (3–4) mm × ca. 1 mm, midvein near upper margin of blade, very unequally sided, base obliquely truncate, apex acute, mucronate. Flowers supra-axillary, mostly solitary, sometimes 2 or 3 together in a very short raceme; bracts and bracteoles similar to stipules but latter smaller. Sepals lanceolate, 4–8 mm, apex acute. Petals bright yellow, unequal, obovate to orbicular, equal to or slightly longer than sepals, shortly clawed. Stamens 10, alternately 5 shorter and 5 longer; anthers opening by apical pores. Ovary with stiff, appressed hairs; stigma flat. Legume flat, falcate, (2.5–5) cm × ca. 0.5 cm; stalk 1.5–2 cm long. Seeds 10–20, flat, smooth.

Distribution: Pohnpei

Utilization: It contains high protein, preferred by cattle and sheep and other livestock, and is hence a good plant for forage.

Plant

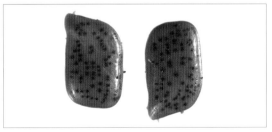

Seeds

Flower

Abrus precatorius

Pohnpei: Kaigus, Kaikes en iak

Lianas. Stem slender, much branched, sparsely white strigose. Leaves paripinnate; petiolule short; leaflets 8-13-paired, opposite; blades suboblong, (1–2) cm × (0.4–0.8) cm, membranous, sparsely white strigose abaxially, glabrous adaxially, rounded at base, truncate and with mucro at apex. Racemes axillary, 3–8 cm. Flowers small, dense. Calyx campanulate, 4-toothed, white strigose. Corolla purple; standard with triangular claw; wings and keels narrower. Stamens 9. Ovary hairy. Legumes oblong, (2–3.5) cm × (0.5–1.5) cm, leathery, dehiscent. Seeds 2–6 smooth, glossy, black in lower part, red in upper part, subglobose.

Distribution: Pohnpei

Utilization: Tender stems and leaves are of good palatability and can be used as forage. Seeds are poisonous and can not be used.

Inflorescence

Plant

Fruit pods

Alysicarpus vaginalis

English: Alyce clover, Alysicarpus, Buffalo clover

Herbs, perennial. Stem erect or procumbent, 30–90 cm tall, glabrous or slightly pubescent. Leaves 1-foliolate; petiole 5–14 mm, glabrous; blade often ovate-oblong or oblong-lanceolate to lanceolate, 6.5 cm × (1–2) cm on upper stem, cordate, nearly orbicular, or ovate, (1–3) × ca. 1 cm on lower stem, abaxially slightly pubescent, adaxially glabrous. Racemes axillary or terminal, 1.5–7 cm, 6–12 - flowered, binate at each node; internodes 2–5 mm. Pedicel 3–4 mm. Calyx membranous, 5–6 mm, 5- lobed, slightly longer than first article of legume. Corolla red, reddish purple, purplish blue, or yellow, slightly longer than calyx, ca. 5 mm; standard obovate. Ovary pubescent, 4–7- ovuled. Legume compressed, cylindric, (1.5–2.5) cm × (2–2.5) mm, pubescent, 4–7- jointed, not constricted between articles, with raised linear ridges. Seeds ellipsoidal, slightly compressed.

Distribution: Pohnpei, Yapese, Chuuk, Kosrae

Utilization: It is an important companion species in natural grassland. It has good palatability and is one of the leguminous plants with high feeding value.

Plant

Inflorescence

Fruit pods

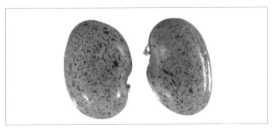

Seeds

Dendrolobium umbellatum

Dwarf shrubs to shrubs or small trees, upto 3 m tall. Young shoots terete, densely appressed yellow or white sericeous; older branches glabrascent. Leaves 3-foliolate; petiole 2–5 cm; petiolule 0.5 – 2 cm long; terminal leaflet blade elliptic or ovate to orbicular or broadly ovate, 5–14 (–17) cm × 3–7(–8.5) cm, lateral leaflets slightly smaller, abaxially appressed long hairy, adaxially glabrescent; lateral veins 7–12 on each side of midvein, reaching margin; stiple filiform or subulate, 1.5–5 mm long. Umbels often 10–20 - flowered, axillary. Pedicel 3–7 mm at anthesis, 5–12 mm in fruit. Bracts ovate, about 2 mm long; Calyx 4–5 mm long, abaxially sericeous, upper lobe 2–toothed at apex. Corolla white; standard broadly obovate or elliptic, (1–1.3) cm × (6–10) mm, clawed; wings narrowly elliptic, (1.1–1.2) cm × (1–2) mm, clawed; keel broader than wings, (1.1–1.2) cm × (3–5) mm, clawed. Stamens ca. 1 cm long. Pistil upto 1.5 cm long; ovary sericeous; style ca. 1.2 cm long, glabrous. Legume narrowly oblong, (2–3.5) cm × (0.4–0.6) cm, (3–)8-jointed; articles broadly elliptic or oblong. Seed elliptic or broadly elliptic, ca. 4 × 3 mm.

Inflorescence

Distribution: Pohnpei, Yapese, Chuuk, Kosrae

Utilization: It is one of the most valuable leguminous forage plants in the Federated States of Micronesia because of its tender branches, abundant leaves and good palatability.

Plant

Infructescence

Tadehagi triquetrum

Shrubs or subshrubs. Stem erect, 1–2 m tall. Shoots 3 ribbed, sparsely short hispid on ribs, glabrescent when old. Leaves 1-foliolate; petiole 1–3 cm, with wing 4–8 mm wide; blade narrowly lanceolate to ovate–lanceolate, (5.8–13) cm × (1.1–3.5) cm, usually more than 3 × as long as wide, abaxially pubescent on midvein and lateral veins, adaxially glabrous, base rounded or shallowly cordate, apex acute or acuminate. Raceme terminal or axillary, 15–30 cm, 2- or 3- flowered at each node. Pedicel 2–6 mm at anthesis, 5–8 mm at fruiting, with spreading, minute, hooked and silky hairs. Calyx broadly campanulate, ca. 3 mm; tube 1.5 mm. Corolla pink to bluish or reddish purple, 5–6 mm; standard nearly orbicular, emarginate at apex; wings obovate, auriculate, clawed; keel arcuate, auriculate, clawed. Stamens diadelphous. Ovary densely puberulent except at upper part of style. Legume 5–8-jointed; articles not reticulate veined, densely yellowish or whitish strigose. Seed transversely broadly elliptic or elliptic, (2–3) mm × (1.5–2.5) mm.

Distribution: Yapese

Utilization: It is a good forage plant preferred by both cattle and sheep.

Plant

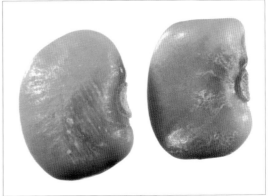
Seeds

Inflorescence

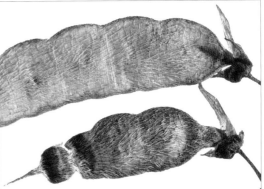
Fruit pods

Desmodium triflorum

Pohnpei: Kamelimel

Herbs, perennial, prostrate, 10–50 cm tall, spreading pubescent. Stems slender, multibranched. Leaves 3-foliolate; petiole ca. 5 mm; terminal leaflet blade obcordate, obtriangular, or obovate, (2.5–10) mm × (2.5–10) mm, base cuneate, apex truncate, slightly emarginate; stiples 0.5–2 mm. Flowers solitary or 2 or 3 in leaf axils. Pedicel 3–8 mm, elongated to 1.3 cm at fruiting. Calyx densely villous, 5-parted; lobes narrowly lanceolate, longer than tube. Corolla purple-red, nearly as long as calyx; standard obcordate, base attenuate, long clawed; wings elliptic, shortly clawed; keel slightly falcate, longer than wings, curved, long clawed. Stamens diadelphous. Pistils about 4 mm; ovary linear. Legume narrowly oblong, slightly falcate, flat, (5–12) mm × ca. 2.5 mm, lower suture undulate, upper suture straight, 3–5 - jointed; articles nearly quadrate, with short, hooked hairs, reticulate veined.

Distribution: Pohnpei, Yapese, Chuuk, Kosrae

Utilization: It is an important companion species in natural grassland, preferred by cattle and sheep and can be used for grazing.

Desmodium heterocarpon

Pohnpei: Kaamalimal, Kamelimel, Kehmelmel en nansapw

Shrubs or subshrubs, erect or prostrate, 30–150 cm tall, much branched from base of stem. Leaves 3-foliolate; petiole 1–2 cm, slightly pubescent; terminal leaflet blade elliptic, narrowly elliptic, or broadly obovate, (2.5–6) cm × (1.3–3) cm, abaxially white appressed pubescent, adaxially glabrous, base obtuse, apex rounded or obtuse, emarginate, mucronate. Lateral leaflet blades usually smaller; stiples filiform, about 5 mm; petiolules 1–2 mm, densely appressed roughly haired. Racemes terminal or axillary, 2.5–7 cm; rachis with white, spreading, hooked hairs or yellowish or white, straight, appressed hairs, densely flowered. Bracts ovate lanceolate, ciliated. Pedicel 3–4 mm. Calyx 1.5–2 mm, campanulate, 4-lobed; upper lobes slightly 2-toothed at apex. Corolla purple, purple-red, or white, ca. 5 mm; standard obovate–oblong, shortly clawed; wings obovate, auriculate, clawed; keel extremely curved, apex obtuse. Stamens diadelphous, about 5 mm; pistils about 6 mm. Infructescence crowded. Legume erect, narrowly oblong, (1.2–2) cm × (2.5–3) mm, upper suture shallowly undulate, both sutures hooked hairy, 4–7-jointed; articles quadrate.

Distribution: Yapese, Kosrae

Utilization: It is a legume with higher feeding value and it can be used for grazing or cultivation as cut-and-carry forage.

Plant

Inflorescence

Fruit pods

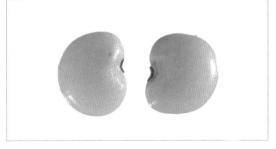
Seeds

Desmodium incanum

English: Kaimi clover, Spanish clover, Tick trefoil

Subshrubs or shrubs, prostrate, ascending, or erect. Stems pubescent with spreading hooked and longer straight hairs, glabrate. Leaves trifoliolate, leaflets subcoriaceous, usually elliptic or narrowly elliptic, terminal one 4–9 cm long, 1.5–4.5 cm wide, lateral nerves conspicuous, upper surface with minute hooked and straight hairs, lower surface densely appressed pubescent, apex obtuse or often acute; lateral leaflets smaller; petioles 1–4 cm long. Flowers numerous in racemose inflorescences, 5–12 cm long, rachis densely pubescent with minute hooked hairs, pedicels 3–10 mm long, persistent after articles fall, pubescent with minute hooked hairs; corolla pink to purplish, 5–6 mm long. Pods stipitate, 2.5–3 cm long, 3–8-jointed, densely pubescent with hooked hairs; articles nearly semicircular, 4–5 mm long, ca. 3 mm wide.

Plant

Distribution: Pohnpei, Yapese, Chuuk, Kosrae

Utilization: It is a legume plant with higher feeding value, and can be used for grazing or as cut-and-carry forage.

Inflorescence

Fruit pods

Desmodium tortuosum

English: Beggarweed, Dixie ticktrefoil, Florida beggarweed

Herbs, perennial, erect, 50–200 cm tall. Stems with hooked hairs, sometimes intermixed with long hairs. Leaves 3-foliolate, rarely 1-foliolate, papery; stipules persistent, lanceolate, 5–8 mm; petioles 1–8 cm, as hairy as stems; leaf rachis 0.5–2 cm; terminal leaflet blade oblong, sometimes rhomboid ovate, 3–8(–14) × 1.5–3(–6) cm, both surfaces sparsely hairy, base cuneate, apex obtuse; lateral leaflet blades mostly ovate, (2.5 – 4) cm × (1–2.5) cm. Racemes or sometimes panicles, terminal or axillary; rachis with dense minute hooked and glandular hairs, 2-flowered at each node. Pedicel filiform, to 1.7 cm at fruiting, hairy as rachis. Calyx 3–4 mm, 5-parted; lobes longer than tube. Corolla red, white, or yellow; standard obovate, (2.5–3.5) mm × ca. 2 mm, base attenuate, apex emarginate; wings oblong, base auriculate, shortly clawed; keel obliquely oblong, clawed. Legume narrowly oblong, 1.5–2 cm, both sutures constricted between articles, moniliform, densely gray-yellow hooked hairy, (3–) 5–7 - jointed.

Distribution: Kosrae

Utilization: It is a legume plant with higher feeding value and it can be used for grazing or as cut-and-carry forage.

Plant

Stipules

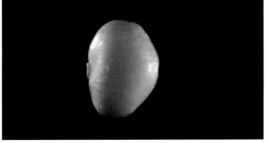

Seed

Desmodium heterophyllum

English: Hetero, Spanish clover, Variable-leaf ticktrefoil

Herbs, prostrate or ascending, 10–70 cm tall. Young parts spreading pubescent. Leaves 3-foliolate, often intermixed with 1-foliolate leaves on lower part; petiole 0.5–1.5 cm, sulcate, sparsely pudescent; terminal leaflet blade broadly elliptic or broadly elliptic-obovate, (1–3) cm × (0.8–1.5) cm, base obtuse, apex rounded or nearly truncate, often emarginate. Flowers solitary or binate in leaf axils or 2 or 3 scattered on rachis. Pedicel 1–2.5 cm. Calyx 5-parted; upper 2 lobes deeply incised near base, villous and minutely hooked hairy. Corolla purple-red to white, ca. 5 mm; standard broadly obovate; wings obovate or narrowly elliptic, shortly auriculate; keel slightly curved, shortly clawed. Stamens diadelphous, about 4 mm. Pistils about 5 mm;.ovary appressed pubescent. Legume narrowly oblong, straight or slightly curved, (12–18) mm × ca. 3 mm, lower suture deeply undulate, upper suture straight, 3–5-jointed, flat; articles broadly oblong or quadrate, 3.5–4 mm, glabrescent, reticulate veined.

Distribution: Pohnpei

Utilization: It is a common companion species in natural grassland, and can be used for grazing.

Plant

Leaves

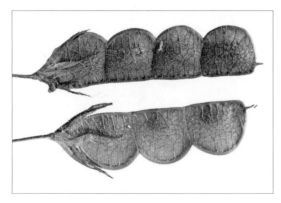
Fruit pods

Seeds

Stylosanthes hamata

English: Caribbean stylo, Caribbean stylo

Herbs or subshrubs, 10–50 cm tall, prostrate or ascending. Stems sparsely pubescent. Stipules sheathing, 4–12 mm. Leaves 3-foliolate; petiole short; petiolules ca. 0.5 mm; stiples absent; leaflet blades ovate, elliptic, or lanceolate, (8–14) mm × (3–5) mm, commonly glabrous, base cuneate, margin setose, apex acute and mucronate. Inflorescences axillary or terminal, 1–1.5 cm, with 2–10 clustered flowers, each flower subtended by a rudimentary axis; primary bracts 1–1.2 cm, spreading setose; secondary bracts (2–3.5) mm × ca. 0.5 mm. Bracteoles ca. 2 mm. Hypanthium 2–2.5 mm. Calyx tube oblong, ca. 2 × 1 mm. Corolla yellow, with red slender striations; standard ca. 4 × 3 mm. Legume quadrate, (2–3.5) mm × ca. 2 mm, commonly pubescent, with 2 articles, beak 3–3.5 mm and uncinate. Seeds light brown, ellipsoid, ca. 2 × 1 mm.

Distribution: Pohnpei, Yapese, Chuuk, Kosrae

Utilization: It is a legume plant with higher feeding value, and it can be used for grazing or as cut-and-carry forage.

Plant

Fruit pods

Inflorescence

Seeds

Aeschynomene indica

Pohnpei: Ikin sihk

Shrublets or annual herbs, 30–100 cm tall. Stems erect, many branched, cylindric, hollow, glabrous, corky at base, often with nodule-bearing adventitious roots. Stipules elliptic to lanceolate, (4–11) cm × 1–2 cm, membranous, caducous, base auriculate, apex acuminate. Leaves 20–60-foliolate, often sensitive; petiole 2–4 mm; rachis with tuberculate-based trichomes; leaflet blades linear-oblong, (3–13) mm × (1–3) mm, papery, base oblique, apex obtuse and mucronate. Inflorescences axillary, racemose, 1,5–2 cm long, sometimes short or reduced to a solitary flower; peduncle 4–7 mm, with tuberculate-based trichomes; bracts ovate, caducous, margin often denticulate. Bracteoles ovate-lanceolate, persistent. Calyx 3–4 mm, membranous, glabrous. Corolla pale yellow with purplish longitudinal striations; standard larger, subrounded, very short clawed at base. Stamens diadelphous; ovary compressed, linear. Legume linear-oblong, (2.2–3.4) cm × (3–5) mm, straight or slightly curved, herbaceous to leathery, abaxial suture straight, slightly indented; articles 2–8, quadrate, slightly muricate and with tuberculate-based trichomes. Seeds blackish brown, reniform, (3–3.5) mm ×(2.5–3) mm.

Distribution: Pohnpei, Yapese

Utilization: It is a legume plant with higher feeding value and can be used for grazing or as cut-and-carry forage.

Plant

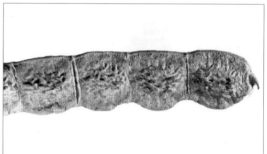

Part of fruit pods

Seeds

Aeschynomene americana

English: American joint vetch

Herbs or shrublets, annual or short-lived perennial, 0.5–2 m tall. Stems erect, 30–50 branched, glabrous, viscid, 3–9 mm in diameter; branches tomentose. Stipules lanceolate, (10–12) mm × (1–3) mm, membranous, base auriculate, apex acute. Leaves 30–40-foliolate; leaflet blades alternate, linear-oblong, (8–10) mm × (2–4) mm, papery, primary veins 2–4, base oblique, apex obtuse and mucronate. Inflorescences axillary, racemose, laxly branched, 2–4-flowered; bracts cordate, membranous. Bracteoles linear–ovate, striate. Calyx deeply 2-lipped. Corolla yellow, ca. 8 mm. Ovary 2-many loculed; Legume oblong, (2–4) cm × (2.5–3) mm, herbaceous to leathery, slightly curved, abaxial suture undulate and indented; articles 4–8, rounded, slightly muricate. Seeds 5–8, brown, reniform, ca. 2 mm × 1 mm.

Distribution: Yapese, Kosrae

Utilization: It a legume plant with higher feeding value and can be used for grazing or as cut-and-carry forage.

Plant parts

Plant cluster

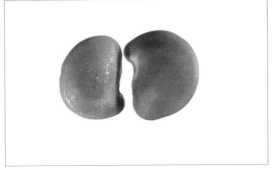
Seeds

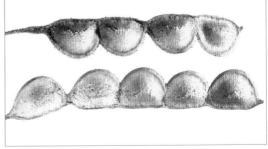
Fruit pods

Arachis hypogaea

Herbs, annual, erect to decumbent. Stems erect or procumbent, 30–80 cm tall, yellowish pubescent, glabrescent. Stipules 2–4 cm, pilose. Leaves usually 4-foliolate; petiole 3.7–10 cm, covered with long flexuous trichomes, basally adnate to stipule; petiolules 1–10 mm, velutinous; leaflet blades ovate-oblong to obovate, (1.1–5.9) cm × (0.5–3.4) cm, papery, both surfaces with long trichomes, veins ca. 10 on each side of midvein, base almost rounded, margin ciliate, apex obtuse or emarginate and mucronate. Bracts lanceolate, apex acuminate. Flowers 8–10 mm, sessile; bracteoles lanceolate, ca. 5 mm, velutinous. Calyx tube 4–6 mm, thin. Corolla yellow to golden yellow; standard spreading, apex emarginate; wings distinct, oblong to obliquely ovate, slender; keels distinct, long ovate, shorter than wings, inflexed, apex acuminate to beaked. Ovary oblong; style longer than calyx; stigma terminal, small, sparsely pubescent. Legume geocarpic, oblong, inflated, (2–5) cm × (1–1.3) cm, thick-walled, reticulate veined, with 1–4(–6) seeds. Seeds light brown, oblong, 5–10 mm in diam.

Distribution: cultivated in Yapese

Utilization: Haulm is favored by livestock such as cattle, sheep, etc. and is of very high feeding value.

Plant

Flemingia macrophylla

Shrubs, erect, 0.8–2.5 m tall. Young branches densely appressed silky villous. Leaves digitately 3-foliolate; stipules lanceolate, up to 2 cm, villous, with glandular striations, apex long acuminate, usually deciduous; petiole 3–6 cm, narrowly winged; petiolules 2–5 mm, densely hairy; leaflets papery to thinly leathery; terminal leaflet broadly lanceolate to elliptic, (8–15) cm × (4–7) cm, glabrous except for veins, abaxial surface with small dark brown sessile glands, basal veins 3, base broadly cuneate, apex acuminate; lateral leaflets smaller, oblique, base rounded at one side, cuneate at other. Racemes usually clustered at axil, 3–8 cm, with many clustered flowers; peduncle usually extremely short. Calyx campanulate, 6–8 mm, shortly very pale brown vil–lous; lobes linear-lanceolate, ca. 2 × as long as tube, lower one longest. Corolla purple, slightly longer than calyx; standard oblong, shortly clawed, auriculate; wings narrowly elliptic, slen-derly clawed, one with auricle; keel oblong, long clawed, apex slightly curved. Ovary elliptic, with very pale brown hairs. Legume elliptic, (10–16) mm × (7–9) mm, sparsely pubescent, apex with small acute beak. Seeds 1 or 2, glossy black, orbicular.

Distribution: Yapese, Kosrae

Utilization: It is a legume plant with higher feeding value and can be used for grazing or as cut-and-carry forage.

Plant

Pods

Inflorescence

Seeds

Flemingia strobilifera

English: Luck plant, Wild hops

Shrubs, 0.3–3 m tall. Branchlets ribbed, densely gray to dull brown villous. Leaves simple; stipules linear-lanceolate, 0.8–1.8 cm, persistent or deciduous; petiole usually 0.3–1.5 cm, densely hairy; leaf blade ovate, narrowly ovate, ovate-elliptic, broadly elliptic, or oblong, (6–15) cm × (3–7) cm, thinly leathery, glabrous or almost glabrous except for veins, lateral veins 5–9 pairs, base rounded, slightly cordate, apex acuminate, obtuse, or acute. Inflorescence a thyrse, sometimes branched, 5–11 cm; inflorescence rachis densely gray brown villous; cymules each enclosed by concave bract; bracts (1.2–3) cm ×

Plant

(2–4.4) cm, papery to almost leathery, both surfaces long hirsute, margin ciliate, apex truncate or rounded, slightly emarginate and with slender mucro. Flowers small; pedicel 1.5–3 mm. Calyx pubescent; lobes ovate, slightly longer than tube. Corolla longer than calyx; standard broadly orbicular; wings narrower than keels. Legume elliptic, (6–10) mm × (4–5) mm, sparsely pubescent, inflated. Seeds 2.

Distribution: Yapese

Utilization: It is a legume plant with higher feeding value and can be used for grazing or as cut-and-carry forage.

Leaves

Flower

Crotalaria pallida

Chuuk: Afalafal

Pohnpei: Kandalahria, Klodalahria wah tikitik, Krodalaria

Herbs, perennial. Branches terete, ribbed, densely appressed pubescent. Stipules acicular, very minute, caducous. Leaves 3-foliolate; petiole 2–4 cm; petiolules 1–2 mm; leaflet blades oblong to elliptic, (3–6) cm × (1.5–3) cm, abaxially sparsely sericeous, adaxially glabrous, veins distinct on both surfaces, base broadly cuneate, apex obtuse to retuse. Racemes terminal, ca. 25 cm, 10–40-flowered; bracts linear, ca. 4 mm, caducous. Pedicel 3–5 mm; bracteoles inserted at base of calyx tube, similar to bracts, ca. 2 mm. Calyx subcampanulate, 4–6 mm, 5-lobed, densely pubescent; lobes triangular, ± as long as tube. Corolla yellow, exserted beyond calyx; standard orbicular to elliptic, ca. 1 cm in diam., base with 2 appendages; wings oblong, ca. 8 mm, marginally pilose on basal part; keel ca. 1.2 cm, rather shallowly rounded, marginally pilose at base, beak narrow and ± projecting. Ovary subsessile. Legume oblong, (3–4) cm × (0.5–0.8) cm, 20–30-seeded, pilose when young but glabrescent; carpels twisted at dehiscence.

Seeds

Distribution: Pohnpei, Yapese, Chuuk, Kosrae

Utilization: It is usually used as green manure.

Plant

Inflorescence

Pod

Crotalaria incana

English: Fuzzy rattlebox, Fuzzy rattlepod

Herbs, upto 1 m tall. Stems brownish yellow spreading pubescent. Stipules acicular, 2–3 mm, tardily caducous. Leaves 3-foliolate; petiole 3–5 cm; petiolules 1–3 mm; leaflet blades elliptic–obovate, obovate, or suborbicular, (2–4) cm × (1–2) cm, terminal one larger than lateral ones, thin, abaxially pubescent to subglabrous, adaxially glabrous, secondary veins 6–10 on each side of midvein, secondary and tertiary veins abaxially distinct and adaxially inconspicuous, base rounded to broadly cuneate, apex obtuse and mucronate. Racemes terminal or leaf-opposed, 10–20 cm, 5–15-flowered; bracts 1–10 mm, caducous. Pedicel 3–4 mm; bracteoles inserted at base of calyx tube, similar to bracts, 2–3 mm. Calyx subcampanulate, 6–8 mm, 5-lobed, pubescent; lobes lanceolate, longer than tube. Co-rolla yellow, exserted beyond calyx; standard elliptic, 8–10 mm, base with appendages, apically usually clustered pilose; wings oblong, 8–10 mm; keel about equal to wings, abruptly rounded below middle, beak well developed and incurved. Legume clavate, (2–3) cm × (0.7–1) cm, apically slightly oblique, 20–30-seeded, densely rusty pilose; stipe ca. 2 mm.

Distribution: Pohnpei

Utilization: It is usually used as green manure.

Plant

Pod

Inflorescence

Crotalaria assamica

Engllish: Indian rattlebox

Herbs, erect or ascending, upto 1.5 m tall. Branches terete, sericeous. Stipules linear, minute. Leaves simple; petiole 2–3 mm; leaf blade oblanceolate to narrowly elliptic, (5–15) cm × (2–4) cm, thin, abaxially rusty puberscent, adaxially glabrous, base cuneate, apex obtuse and mucronate. Racemes terminal or leaf-opposed, to 30 cm, 20–30-flowered; bracts linear, 1–2 mm. Bracteoles similar to bracts but shorter. Calyx 2-lipped, 1–1.5 cm, pubescent; lobes lanceolate-triangular, equal to tube. Corolla golden yellow; standard suborbicular to elliptic, 1.5–2 cm, base with 2 appendages, apex retuse; wings 1.5–1.8 cm; keel rounded through 90°, narrowed apically from middle and extended into a long twisted beak exserted beyond calyx. Ovary glabrous. Legume oblong, (4–6) cm × ca. 1.5 cm, 20–30-seeded; stipe ca. 5 mm.

Seeds

Distribution: Kosrae

Utilization: It is usually used as a green manure.

Plant

Inflorescence

Pods

Kummerowia striata

Herbs, diffuse or prostrate. Stem and branchlets with downward-pointing white hairs. Stipules ovate-oblong, 3–4 mm, longer than petiole, striate, long ciliate; petiole 1–2 mm; leaflets obovate, narrowly obovate, or oblong, terminal one (0.6–2.2) cm × (3–8) mm, lateral veins dense, base nearly rounded or broadly cuneate, apex rounded, rarely emarginate. Flowers 1–3 in upper axils of leaves. Pedicel ca. 1 cm, glabrous; bracteoles 4, attached to proximal part of calyx, 1 very small bracteole placed at article of pedicel, others larger. Calyx campanulate, tinged with purple, 5-lobed; lobes broad ovate. Corolla pink or purple, 5–6 mm; standard elliptic, base attenuate, clawed, auriculate; keel subequal to or slightly longer than standard; wings slightly shorter than keel. Legume orbicular or obovoid, slightly compressed, 3.5–5 mm, ca. 2 times longer or slightly longer than calyx, small pubescent, apex mucronate.

Distribution: Pohnpei, Yapese, Chuuk, Kosrae

Utilization: It a legume plant with higher feeding value and can be used for grazing.

Plant cluster

Leaves

Inflorescence

Pods

Cajanus cajan

English: Congo pea, Pigeon pea, Red gram

Shrubs, erect, 1–3 m tall. Branchlets gray pubescent. Leaves pinnately 3-foliolate; stipules small, ovate-lanceolate, 0.2–3 mm; petiole 1.5–5 cm, sparsely pubescent; stipels extremely small; petiolules 1–5 mm, hairy; leaflets lanceolate to elliptic, (2.8–10) cm × (0.5–3.5) cm, papery, abaxial surface densely pubescent and with inconspicuous yellow glands, adaxial surface pubescent, apex acute or acuminate, usually mucronate. Raceme 3–7 cm; peduncle 2–4 cm; a few flowers terminal or almost terminal; bracts ovate-elliptic. Calyx campanulate, 5–7 mm; lobes triangular or lanceolate, pubescent. Corolla yellow, ca. 3 times longer than calyx; standard suborbicular, with inflexed auricle; wings slightly obovate, with short auricle; keel obtuse at apex, slightly inflexed. Stamens diadelphous. Ovary hairy; ovules several; style long, linear, glabrous; stigma capitate. Legume linear-oblong, (4–8.5) cm × (0.6–1.2) cm, dun pubescent, apex beaked, acuminate, long mucronate. Seeds 3–6, gray, sometimes with brown spots, subspherical, ca. 5 mm in diam., slightly compressed; strophiole absent.

Distribution: Cultivated in Yapese

Utilization: It is a legume plant with higher feeding value and can be used for grazing or as cut-and-carry forage.

Inflorescence

Plant

Seeds

Macroptilium atropurpureum

English: Purple bushbean, Purple-bean

Perennial prostrate herbs, sometimes rooting at nodes. Stems pubescent or tomentose. Stipules ovate, 4–5 mm, pilose; petiole 0.5–5 cm; leaflets ovate to rhombic, (1.5–7) cm × (1.3–5) cm, sometimes lobed, lateral ones oblique, lobed on outer side, abaxially silvery tomentose, adaxially pubescent, base rounded, apex obtuse or acute. Inflorescences with peduncle 10–25 cm and rachis 1–8 cm. Calyx campanulate, ca. 6 mm, white pilose. Corolla deeply blackish purple; standard 1.5–2 cm, with long claw. Legumes linear, (5–9) cm × ca. 4 mm, apex rostrate, 12–15-seeded. Seeds marbled with brown and black striae, oblong-elliptic, ca. 4 mm.

Distribution: Pohnpei, Yapese

Utilization: It is a legume plant with higher feeding value and can be used for grazing or as cut-and-carry forage.

Plant cluster

Flowers

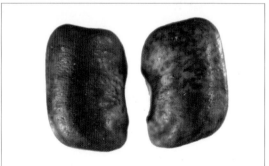

Seeds

Lablab purpureus

English: Bonavist, Bonavist-bean, Dolichos

Herbs, twining. Stems up to 6 m, usually purplish. Stipules lanceolate; leaflets deltoid-ovate, (6–10) cm × (6–10) cm, lateral ones oblique, base subtruncate, apex acute or acuminate. Racemes axillary, erect, 15–25 cm. Bracteoles 2, suborbicular, 3 mm long, deciduous; Flowers 2–5 clustered at each node. Calyx ca. 6 mm, upper 2 teeth wholly connate, lower 3 subequal. Corolla white or purple; standard orbicular, ca. 12 mm, auriculate appendaged at both sides of base; wings with blade ca. 10 mm, broad obovate, with a truncate auricle; keel base attenuate, clawed. Ovary linear; style longer than ovary. Legumes oblong-falcate, (5–7) cm × (1.4–1.8) cm, compressed, straight or slightly curved, beaked. Seeds 3–5, white, purple, or purple-black, oblong; hilum linear.

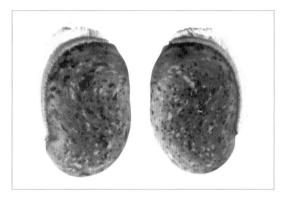

Seeds

Distribution: Cultivated in Yapese

Utilization: It is a legume plant with higher feeding value and can be used for cut-and-carry forage.

Plant

Flowers and pods

Psophocarpus tetragonolobus

Herbs. Stems 2–3 m or more, glabrous. Stipules dorsifixed, ovate to lanceolate, 0.8–1.2 cm; petiole sulcate; leaflets ovate-deltoid, (4–15) cm × (3.5–12) cm, base truncate or rounded, apex acute or acuminate. Racemes axillary, 1–10 cm, 2–12-flowered; peduncles 5–15 cm; bracteoles suborbicular, 2.5–4.5 mm in diam. Calyx campanulate, ca. 1.5 cm. Standard green outside, pale blue inside, ca. 3.5 cm in diam., base appendaged, apex emarginate; wings obovate, pale blue, ca. 3 cm, with T-shaped auricle at middle of claw; keel white tinged with pale blue, slightly incurved with rounded auricle at base. Legumes yellow-green or green, sometimes with red spots, tetragonal, [10–25(–40)] cm × (2–3.5) cm, wings 0.3–1 cm wide with serrate margins. Seeds 8–17, white, yellow, brown, black, or variegated, 0.6–1 cm in diam, shining, margin arillate.

Distribution: Cultivated in Yapese

Utilization: Its haulm has a high feeding value.

Plant

Inflorescence, pods and flowers

Seed

Vigna unguiculata

Annual herbs, erect, trailing, or twining. Stems 1–3 m, subglabrous. Stipules lanceolate, ca. 1 cm, with a narrow spur below point of attachment; leaflets ovate-rhomboid, (5–15) cm × (4–6) cm, lateral ones oblique, puberulent or glabrous on both surfaces, base acute to rounded, apex acute. Racemes axillary, with 2–6 flowers clustered at top of rachis. Calyx campanulate, 6–10 mm; teeth lanceolate. Standard yellowish white or violet, suborbicular, (1.2–3.3) cm × (1–3.2) cm, apex emarginate; wings blue to purple, subdeltoid; keel usually white or pale, not twisted. Legumes terete, [7.5–30 (–90)] cm × (0.6–1) cm. Seeds several, dark red or black, mottled with black or brown, oblong or reniform, 6–12 mm.

Distribution: Cultivated in Yapese and Kosrae

Utilization: Its haulm has a high feeding value.

Flower

Pod

Vigna marina

Perennial herbs, prostrate or climbing. Stems up to several meters, glabrescent when old. Stipules 2-lobed at base, ovate, 3–5 mm; leaflets ovate-orbicular or obovate, (3.5–9.5) cm × (2.5–7.5) cm, very shortly setose to subglabrous on both surfaces, base broadly cuneate or narrowly rounded, apex rounded, obtuse, or emarginate. Racemes axillary, 2–4 cm, pubescent; peduncles 3–13 cm, sometimes thickened. Bracteoles lanceolate, 1.5 mm long, caduceus. Calyx tube 2.5–3 mm, glabrous; teeth deltoid, 1–1.5 mm long, upper 2 connate and entire, ciliate. Corolla yellow; standard broadly obovate, (1.2–1.3) cm × ca. 1.4 cm; wings and keel ca. 1 cm. Legumes linear-oblong, (3.5–6) cm × (0.8–0.9) cm, pubescent when young, glabrascent, slightly constricted between seeds. Seeds 2–6, yellow-brown or red-brown, oblong, (5–7) mm × (4.5–5) mm; hilum oblong, slightly narrow to one end; testa slightly raised around hilum.

Distribution: Pohnpei, Yapese, Chuuk, Kosrae

Utilization: It is a legume plant with higher feeding value and can be used for grazing or as cut-and-carry forage.

Plant cluster

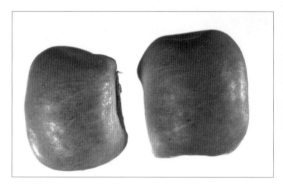
Seeds

Flowers and pods

Clitoria ternatea

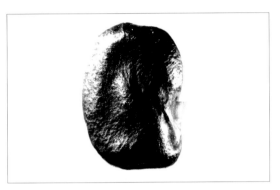

Seed

Herbs. Stems twining, slender, densely deciduous appressed shortly villous. Leaves 2.5–5 cm, pinnately 5–7-foliolate, usually 5-foliolate; stipules small, linear, 2–5 mm; petiole 1.5–3 cm; stipels small, bristlelike; petiolules 1–2 mm; leaflets broadly elliptic or almost ovate, (2.5–5) cm × (1.5–3.5) cm, thinly papery or almost membranous, appressed shortly villous or sometimes glabrous on both surfaces, lateral veins 4 or 5 pairs, base obtuse, apex obtuse, slightly emarginate, usually with mucro. Flowers large, solitary in axil; bracteoles green, small, suborbicular or obovate, membranous, with obvious reticulate veins. Calyx membranous, 1.5–2 cm, 5-lobed; lobes lanceolate, less than 1/2 of tube, apex acuminate. Corolla sky blue, pink, or white, to 5.5 cm; standard faintly white or orange in middle, broadly obovate, ca. 3 cm, base shortly clawed; wings and keels much shorter than standard, both clawed; wings obovate-ob-long; keels elliptic. Ovary villous. Legume brown, linear-oblong, (5–11) cm × (0.7–1) cm, compressed, with long beak. Seeds 6–10, black, oblong, ca. 0.6 × 0.4 cm, with obvious strophiole.

Distribution: Cultivated in Yapese

Utilization: It is usually used as an ornamental plant or as forage.

Plant

Flower

Canavalia cathartica

Herbs, biennial, robust, twining. Stems and branches sparsely pubescent. Stipules small, callose; stipels minute, early caducous. Leaflets ovate, (6–10) cm × (4–9) cm, sparsely white pubescent on both surfaces, base broadly cuneate, truncate, or rounded, apex acute or rounded. Petioles 3–8 cm long; petiolules 5–6 mm, villous. Racemes with 1–3 flowers at each node of rachis. Calyx campanulate, ca. 12 mm, pubescent, upper lip with rounded lobes, shorter than tube, lower lip with 3 teeth. Corolla pink or purplish, 2–2.5 cm; standard orbicular, ca. 2 × 2.5 cm, with 2 thickenings near base, clawed, apex emarginate; wings and keel curved, ca. 2 cm. Ovary villous; style glabrous. Legumes oblong, (7–9) cm × (3.5–4.5) cm, turgid, apex rostrate. Seeds brownish black, elliptic, ca. 18 mm × 12 mm, hard and smooth; hilum 10–14 mm.

Distribution: The outer island of Yapese

Utilization: It is a legume plant with higher feeding value and can be used for grazing or as cut-and-carry forage.

Plant parts

Pods

Calopogonium mucunoides

Pohnpei: Kalopo

Yapese: Hulip nuop

Herbs, procumbent, densely hirsute. Stipules triangular-lanceolate, 4–5 mm; petiole 4–12 cm; stipels subulate; terminal leaflet ovate-rhombic, lateral ones obliquely ovate, (4–10) cm × (2–5) cm, broadly cuneate to rounded at base, acute or obtuse at apex. Inflorescences 1–10 cm; flowers 5 or 6, clustered at nodes of rachis; bracts and bracteoles linear-lanceolate, ca. 5 mm. Calyx tube glabrescent; lobes longer than tube, linear-lanceolate, densely appressed hirsute, long acuminate at apex. Corolla purplish; wings obovate-oblong; keel straight, with short auricles. Anther rounded. Ovary densely hirsute, 5–6 loculed. Legumes linear-oblong, (2–4) cm × ca. 4 mm, with appressed long brown bristles. Seeds 5–6, ca. 2.5 × 2 mm.

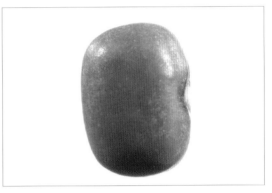
Seed

Distribution: Pohnpei, Yapese

Utilization: It is a legume plant with higher feeding value and can be used for grazing or as cut-and-carry forage or green manure.

Plant cluster

Pods

Sesbania grandiflora

Pohnpei: Pakphul

Small trees, 4–10 m tall. Leaves 20–40 cm, 20–60-foliolate; petiolules 1–2 mm; stipels acerose; leaflet blades oblong, (2–5) cm × (0.8–1.6) cm, smaller at both ends of rachis than in middle, both surfaces with dense appressed purplish brown glands and appressed villous but glabrescent, secondary veins. Ra–cemes 4–7 cm, pendulous, 2–4-flowered. Flowers 7–10 cm, conspicuously falcately curved in bud. Pedicel 1–2 cm, densely appressed pilose. Calyx green, campanulate, (1.8–2.9) cm × (1.5–2) cm. Corolla white, pink, or rosy; standard oblong-obovate to broadly ovate, (5–7.5) cm × (3.5–5) cm, reflexed at anthesis, callus absent, claw ca. 1.6 cm, base subcordate, apex retuse; wings fal-cately long ovate, asymmetric, ca. 5 × 2 cm, claw ca. 2 cm, apex obtuse; keel curved, ca. 5 cm, claw ca. 2 cm, limbs with basal abaxial edges connate, apical 1/4–1/3 free, apex obtuse. Stamens ca. 9 mm; anthers linear, 4–5 mm, dorsifixed. Pistil linear, ca. 8 cm, compressed, falcately curved, glabrous; ovary stipitate; stigma slightly turgid. Legume linear, slightly curved, nodding, (20–60) cm × (7–8) mm, ca. 8 mm thick, dehiscent, car-popodium ca. 5 cm, suture angulate at maturity, apex tapering into a 3–4 cm beak. Seeds reddish brown, ellipsoid to subreni-form, ca. 6 × (3–4) mm.

Distribution: Cultivated in Pognpei and Yapese

Utilization: It is usually grown as an ornamental plant, and its foliage can be used as green manure or forage.

Plant

Flowers

Sesbania cannabina

Herbs, 3–3.5 m tall. Stems green. Leaves 40–60-foliolate; rachis 15–25 cm; petiolules ca. 1 mm, with sparsely appressed trichomes; stipels subulate, sub-equal to petiolules; leaflet blades opposite, (8–20) mm × (2.5–4) mm. Racemes 3–10 cm, 2–6-flowered; peduncle slender, pendulous, sparsely appressed villous; bract linear-lanceolate, caducous. Pedicel slender, pendulous, sparsely appressed villous; bracteoles 2, caducous. Calyx obliquely campanulate, 3–4 mm, glabrous; teeth triangular, with 1–3 appendages between each, inner margin white slender pi-lose, apex acute. Corolla yellow; standard lamina transversely ovate to suborbicular, 9–10 mm, with a ca. 2 mm claw, base subrounded and with a small pyriform callus, apex retuse to rounded; wings obovate-oblong, ± as long as standard, ca. 3.5 mm wide, with transverse corrugation, base shortly auriculate, middle with dark grayish brown spots; keel broadly triangular-ovate, shorter than wings, as long as wide, with a ca. 4.5 mm claw, apex obtuse. Anthers ovate to oblong. Pistil glabrous; stigma capitate. Legume long terete, slightly curved, (12–22) cm × (2.5–3.5) mm, dehiscent, outside with dark brown stripes, trabeculate between seeds, carpopodium ca. 5 mm, apex acute and with a 5–7 (–10) mm beak. Seeds 20–35 per legume, greenish brown, terete, ca. 4 mm × (2–3) mm, glossy; hilum rounded, slightly oblique to one end.

Distribution: Pohnpei, Yapese, Chuuk, Kosrae

Utilization: It is usually used as green manure and occasionally fed on by livestock.

Seeds

Plant

Inflorescence

Sporobolus fertilis

Perennial. Culms densely tufted, erect, rigid, 25–100 cm tall. Leaf sheaths glabrous but margin ciliolate, basal sheaths papery, lightly keeled; leaf blades linear, flat or involute, [15–50(–65)] cm × (0.2–0.5) cm, glabrous or adaxial surface thinly pilose, tapering to a long filiform apex; ligule very short, ca. 0.5 mm, ciliolate. Panicle linear, contracted to spikelike seed-heads, often interrupted especially at base, (7–45) cm × (0.5–1.5) cm; branches 1–2.5(–5) cm, erect and appressed or oblique to main axis, or looser and narrowly ascending, densely spiculate throughout. Spikelets grayish or yellowish green, 1.7–2 mm; lower glume oblong, ca. 0.5 mm, veinless, apex truncate-erose; upper glume oblong-elliptic, 1/2–2/3 spikelet length, 1-veined, ± acute; lemma ovate, as long as spikelet, 1 midribbed, 2 indistinctly lateral-veined, acute. Anthers 3, 0.8–1 mm, yellow. Grain red-brown, obovate-elliptic, 0.9–1.2 mm, distinctly shorter than its lemma and palea, these gaping widely beyond its top, apex truncate.

Distribution: Yapese

Utilization: It is a common plant in natural grassland and usually forms small populations of higher density. It is a grassy forage plant with good palatability and is suitable for grazing.

Plant

Inflorescence

Eleusine indica

Pohnpei: Rehtakai, Resahkai

Yapese: Gucurum

Annual. Culms tufted, erect or geniculate at base, 10–90 cm tall. Leaf sheaths glabrous or tuberculate-pilose, lax, laterally compressed and keeled; leaf blades flat or folded, (10–15) cm × (0.3–0.5) cm, glabrous or adaxial surface tuberculate-pilose; ligule ca. 1 mm, membranous, at most sparsely ciliolate. Inflorescence digitate, spikes (1–)2–7, linear, ascending, (3–10) cm × (0.3–0.5) cm, one spike often set below the rest. Spikelets elliptic, 4–7 mm, florets 3–9; glumes lanceolate, scabrid along keel; lower glume 1-veined, 1.5–2 mm; upper glume with small additional veins in the thickened keel, 2–3 mm; lemmas ovate, 2–4 mm, keel with small additional veins, acute; palea keels winged, shorter than lemma. Grain blackish, oblong or ovate, obliquely striate with fine close lines running vertically between the striae.

Distribution: Pohnpei, Yapese, Chuuk, Kosrae

Utilization: It is a common plant in natural grassland, and suitable for grazing.

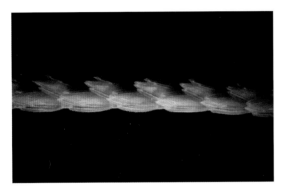

Part of the inflorescence

Plant

Inflorescence

Arundo donax

English: Bamboo reed, Giant reed

Perennial herbaceous grass. Robust reed from a thick knotty rhizome. Culms very stout, erect, 2–6 m tall, 1–1.5 cm in diam., unbranched or with bamboolike clusters of slender branches from nodes. Leaf sheaths longer than internodes, usually glabrous except long pilose at mouth; leaf blades (30–60) cm × (2–5) cm, margins scabrous, tapering to a slender filiform apex; ligule 0.7–1.5 mm. Panicle 30–60 cm, dense, usually purplish; branches 10–25 cm, ascending. Spikelets 10–15 mm, florets 2–5; glumes narrowly lanceolate, 8–12 mm, 3–5-veined, lower glume acute, upper glume sharply acuminate; lemmas linear-lanceolate, 8–11 mm, 3–7-veined, dorsal hairs 5–6 mm, apex minutely bidentate with 1–2 mm awnlet from sinus, lateral veins also shortly extended; palea 1/2 length of lemma body. Stamens 3. Grain minute, black.

Spikelet

Distribution: Pohnpei, Kosrae

Utilization: Cattle like to feed on its young branches and leaves.

Plant

Eragrostis atrovirens

English: Thalia love grass

Perennial grass. Culms loosely tufted, erect or geniculate at base, 15–100 cm tall, ca. 4 mm in diam., 4–8-noded. Leaf sheaths glabrous but pilose along summit; ligules a ciliolate membrane, 0.2–0.3 mm; leaf blades flat or involute, (4–17) cm × (0.2–0.4) cm, adaxial surface scabrous, near base pilose, abaxial surface glabrous. Panicle open, (5–20) cm × (2–4) cm; branches one to several per node. Spikelets plumbeous and purplish, narrowly oblong, (5–15) mm × (1.5–2.5) mm, 8–40-flowered, pedicels 0.5–5 mm; rachilla persistent. Glumes 1-veined, 1–2.3 mm; lower glume ovate, 1–1.3 mm, apex acute, upper glume narrowly ovate, 1.3–2.3 mm, apex acuminate. Lemmas broad ovate, 1.8–2.2 mm, apex acute, lower lemma 2–2.2 mm, deciduous with palea. Palea loosely ciliate along keel, 1.6–1.8 mm. Stamens 3; anthers 0.7–0.9 mm. Caryopsis ca. 1 mm.

Spikelet

Distribution: Pohnpei, Yapese

Utilization: It is a common plant in natural grassland and usually forms small populations of higher density. It is a grassy forage plant with good palatability and is suitable for grazing.

Plant cluster

Inflorescence

Eragrostis unioloides

English: Chinese lovegrass

perennial. Culms erect or geniculate at base, 20–60 cm tall, 2–3 mm in diam., 3–5-noded. Leaf sheaths glabrous and smooth, long pilose along the summit; ligules membranous, ca. 0.8 mm; leaf blades sublanceolate, flat, (2–20) cm × (0.3–0.6) cm, adaxial surface long pilose, abaxial surface smooth, apex acuminate. Panicle open, oblong, (5–20) cm × (3–5) cm; branch solitary, glabrous in axils. Spikelets purplish red at maturity, oblong, (5–10) mm × (2–4) mm, with pedicel 0.2–1 cm, 10–20-flowered; florets closely imbricate; rachilla persistent; glume lanceolate; lower glume 1.5–2 mm, upper glume 2–2.5 mm. Lemmas broadly ovate, veins prominent, apex acute, the lower lemma ca. 2 mm. Palea slightly shorter than lemma, 2-keeled, very narrowly winged and ciliolate, falling off together with its lemma at maturity. Stamens 2; anthers purple, 0.2–0.5 mm. Caryopsis compressed, ellipsoidal, ca. 0.8 mm.

Spikelet

Distribution: Kosrae

Utilization: It is a common plant in moist grasslands and usually forms small populations of higher density. It is a grassy forage plant with good palatability and is suitable for grazing.

Plant cluster

Inflorescence

Centotheca lappacea

Pognpei: Reh

Perennial from a knotty base. Culms solitary or loosely tufted, erect, smooth, 40–100 cm tall, 4–7-noded. Leaf sheaths smooth or ciliate along one margin; leaf blades broadly lanceolate, (5–15) cm × (1–2.5) cm, abaxial surface glabrous with cross veins, adaxial surface glabrous or loosely hispidulous, apex long-attenuate; ligule 1–1.5 mm. Panicle open, 12–25 cm, primary branches up to 15 cm, the spikelets clustered around them; pedicels 2.5–3 mm, slender, pubescent. Spikelets ca. 5 mm, florets 2–3; glumes 3–5-veined; lower glume 2–2.5 mm, acute at apex; upper glume 3–3.5 mm, mucronate at apex; lowest lemma ca. 4 mm, 7-veined, glabrous, apiculate at apex; second and third lemmas 3–3.5 mm, coarsely setose with tubercle-based, reflexing bristles near upper margins; paleas firm, ciliolate along keels. Anthers about 1 mm. Caryopsis ellipsoid, 1–1.2 mm; embryo 1/3 of the length of caryopsis.

Distribution: Pohnpei, Yapese, Chuuk, Kosrae

Utilization: It is gramineous forage plant with good palatability and is suitable for grazing or as cut-and-carry forage.

Plant

Spikelet

Inflorescence

Zoysia japonica

English: Japanese grass

Perennial grass, with long slender stolons, forming large mats. Culms erect, up to 20 cm tall, sometimes branched at base. Leaf sheaths glabrous, pilose at mouth with 1–2 mm hairs, basal sheaths persistent; leaf blades aggregated toward culm base, linear-lanceolate, flat or margins involute, tough, patent, 2.5–6 cm long, 2–4 mm wide, abaxial surface subglabrous, adaxial surface thinly pilose, apex pungent. Raceme spiciform, linear-elliptic, (2–4) cm × (0.3–0.5) cm, long exserted above leaves; pedicels slender, slightly flexuous, longer than spikelet, up to 5 mm. Spikelets many, loosely overlapping, (2.5–3.5) mm × (1–1.5) mm, yellowish green becoming purplish brown; lower glume absent; upper glume obliquely ovate, obscurely 5–7-veined, upper margins broad, papery, apex obtuse, sometimes mucronate; lemma boat-shaped, slightly shorter than glume, 1-veined; palea absent. Anthers ca. 1.5 mm. Caryopsis 1.5–2 mm.

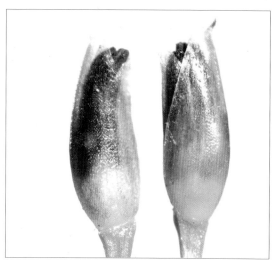
Spikelet

Distribution: Pohnpei, Yapese, Chuuk, Kosrae

Utilization: It is usually used as turfgrass.

Lawn

Plant cluster

Zoysia matrella

Chuuk: Fatil

English: Manila grass

Perennial grass, stoloniferous, mat-forming, also with shallow underground rhizomes. Culms up to 20 cm tall. Leaf sheaths glabrous, bearded at mouth with 4–5 mm hairs; leaf blades flat or involute, tough, suberect to spreading, 3–8 cm long, 1.5–2.5 mm wide, glabrous or adaxial surface thinly pilose, apex acute. Inflorescence linear, (2–4) cm × (0.2–0.3) cm, exserted above leaves; spikelets 10–30, loosely overlapping; rachis somewhat wavy; pedicels shorter than spikelet, 1–3 mm, widened at apex. Spikelets 2–3 × ca. 1 mm, yellowish brown or purplish brown; lower glume usually absent; upper glume lanceolate, 5-veined, midrib prominent, sometimes scabrous toward apex, lateral veins obscure, apex obtuse; lemma membranous, oblong-ovate, 2–2.5 mm, obscurely 3-veined, midvein sometimes shortly excurrent; palea lanceolate, 1/2 as long as lemma. Anthers 1–1.5 mm. Caryopsis ca. 1.5 mm.

Distribution: Pohnpei, Yapese, Chuuk, Kosrae

Utilization: It is usually used as turfgrass and can also be used as forage.

Plant cluster

Spikelet

Chloris formosana

Annual or short-lived perennial grass. Culms erect to decumbent, rooting at lower nodes, 20–70 cm tall. Leaf sheaths laterally compressed, dorsally keeled, glabrous; leaf blades usually folded, 4–40 cm long, 2–3 mm wide, glabrous, apex acute; ligule 0.5–1 mm, ciliate. Racemes digitate, 4–11, erect or somewhat lax, 3–8 cm, pallid or purplish; rachis puberulous. Spikelets with 3 florets, 3-awned; lower glume 1–2 mm; upper glume 2–3 mm, obtuse, mucronate; lemma of fertile floret elliptic in side view, 2.3–3 mm, with a lateral groove, this occasionally appressed pilose, glabrous on keel, densely ciliate on upper margins with ca. 1 mm hairs; awn 4–6 mm; upper florets sterile, lemmas empty, flattened or only slightly inflated, overlapping to form a knob at side of fertile floret; second lemma oblanceolate, truncate, 1.6–2 mm, glabrous, awn 2.5–5 mm; third lemma similar to second but slightly smaller, awn 2–3 mm. Rachilla between infertile florets 0.6–0.7 mm, visible. Caryopsis spindle-shaped, about 2 mm long.

Inflorescence

Distribution: Pohnpei, Yapese, Chuuk, Kosrae

Utilization: It is a gramineous forage plant with good palatability, and is suitable for grazing.

Plant

Cynodon dactylon

English: Bermuda grass, Bahama grass

Perennial grass, stoloniferous, also with slender scaly rhizomes, sward forming. Culms slender, 10–40 cm tall. Leaf sheaths bearded at mouth, otherwise glabrous or thinly pilose; leaf blades linear, short and narrow, 1–12 cm, 1–4 mm wide, usually glabrous, apex subacute; ligule a line of hairs. Racemes digitate, 3–6, 2–6 cm, straight or gently curved, rather stiff, spreading; spikelets overlapping by 1/2–2/3 their length. Spikelets 2–2.7 mm; rachilla extension ca. 1 mm, sometimes with minute rudimentary floret at apex; glumes linear-lanceolate, often purplish, usually more than half as long as floret, 1.5–2 mm, 1-veined, keel scabrous, thickened; lemma as long as spikelet, silky villous along keel, hairs straight, otherwise glabrous or lateral veins thinly villous, apex subacute; palea glabrous, keels scaberulous. Anthers purplish, more than 1 mm. Ovary grabrous; stigma purplish red. Caryopsis subterete, scarcely laterally compressed.

Spikelet

Distribution: Pohnpei, Yapese, Chuuk, Kosrae

Utilization: It is a highly adaptable gramineous plant, distributed in natural grassland of various habitats, and has high feeding and turf value.

Plant cluster

Thuarea involuta

Perennial grass. Culms long and creeping, much branched, rooting at nodes, flowering culms up to 20 cm tall. Leaf sheaths loose, imbricate on the short erect shoots, pilose or only ciliate along margins; leaf blades (2–5) cm × (0.3–0.8) cm, usually puberulous on both surfaces; ligule 0.5–1 mm. Inflorescence a terminal raceme, not exserted from the uppermost spathelike leaf sheath; rachis broad and winglike in lower fertile part, narrowed above in staminate part. Spikelets pubescent; staminate spikelet oblong-lanceolate, 3–4 mm; fertile spikelet ovate-lanceolate, 3.5–4.5 mm.

Inflorescence

Distribution: Pohnpei, Yapese, Chuuk, Kosrae

Utilization: Coastal sandy plants with strong creeping habits that often form a wide stretches of natural grasslands. It can be used as a good plant for greening or as a cover plant for sand fixation in coastal sand, and also has a feeding value, which is suitable for grazing or as cut-and-carry forage.

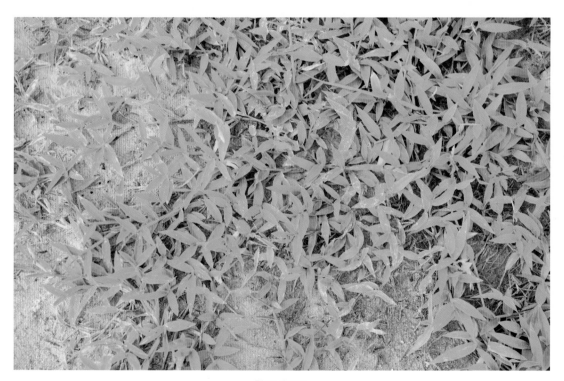

Plant cluster

Setaria geniculata

Yapese: Gatewel

Annual or short-lived perennial with basal buds or a short knotty rhizome. Culms erect or geniculate, 20–90 cm tall. Leaf sheaths keeled, glabrous; leaf blades stiff, flat or involute, (5–30) cm × (0.2–0.8) cm, glabrous or adaxial surface pilose at base, apex acuminate; ligule ca. 1 mm. Panicle densely cylindrical, (2–15) cm × (0.5–1.2) cm; branches reduced to a single mature spikelet subtended by 8–12 bristles; axis pubescent; bristles golden or purplish brown when mature, 2–3 times spikelet length. Spikelets elliptic, 1.8–2.5 mm; lower glume ovate, 1/3 as long as spikelet, acute; upper glume broadly ovate, ca. 1/2 as long as spikelet, obtuse; lower floret neuter; lower palea firmly membranous, lanceolate, about as long as the upper floret but narrower, keels wingless, minutely papillose; upper lemma ovate-elliptic, finely rugose.

Spikelet

Distribution: Yapese

Utilization: It is a graminous forage plant with good palatability, and is suitable for grazing or as cut-and-carry forage.

Population

Pennisetum purpureum

English: Elephant grass, Merker grass, Napier grass

Pohnpei: Poakso, Puk-soh

Perennial grass, forming large tussocks, often with short rhizomes. Culms robust, decumbent, rooting at base, ascending to 2–4 m tall. Leaf sheaths glabrous or hispid; leaf blades linear, up to 120 × 5 cm, abaxial surface glabrous, adaxial surface hispid or papillose-pilose at base, midrib prominent, margins scabrous; ligule 1.5–5 mm. Inflorescence linear, (10–30) cm × (1–3) cm, golden, brownish or purplish; axis densely pilose, closely beset with small peduncle stumps; involucre comprising many slender bristles, enclosing 1–5 spikelets, terminal spikelet fertile, subsessile, laterals when present staminate with 1–2 mm pedicels; inner bristles thinly plumose, longest 1–4 cm. Spikelets 5–7 mm; lower glume vestigial or absent; upper glume 1/4–1/2 as long as spikelet, acute; lower floret staminate or neuter, lemma 1/2–3/4 spikelet length, 5–7-veined, minutely hispidulous, acuminate; upper lemma membranous, obviously 5-veined toward narrowly acuminate tip, lower half cartilaginous, smooth and shiny; anthers with a tuft of short hairs at tip.

Distribution: Chuuk, Kosrae

Utilization: It is of high yield, good quality and good palatability. Domestic animals such as cattle and sheep are fond of feeding on it. It is an excellent graminous forage plant and is suitable for cultivation as cut-and-carry forage.

Blades

Plant cluster

Inflorescence

Pennisetum polystachion

Pohnpei: Poakso

Short-lived perennial or annual grass. Culms much branched, 50–150 cm tall. Leaf blades linear, (10–20) cm × (0.3–1.5) cm, hispid. Inflorescence linear, (10–25) cm × (0.8–1) cm, yellow or purplish; axis angular with sharp decurrent wings below the involucres, these densely packed, often spreading at right angles at maturity; involucre with numerous bristles obscuring the single spikelet, densely ciliate in lower half with crinkled matted hairs, longest bristle 1–2 cm. Spikelet narrowly lanceolate, 3–4.5 mm; lower glume absent or a small triangular scale; upper glume as long as spikelet, 5-veined, obtuse, ciliolate, apiculate; lower floret staminate or neuter, lemma similar but slightly shorter, obtusely 3-lobed; upper floret 2/3 spikelet length, cartilaginous, smooth, shiny, readily deciduous at maturity; anthers without hairs at tip.

Population

Distribution: Pohnpei, Kosrae

Utilization: It is palatable to cattle, sheep and other livestock, and is a high quality graminous forage plant, suitable for grazing.

Inflorescence

Plant

Cenchrus echinatus

English: Bur grass, Burgrass, Burr grass

Annual grass. Culms geniculate, usually rooting at basal nodes, 15–90 cm tall. Leaf sheaths keeled, usually imbricate at base; leaf blades linear or linear-lanceolate, (5–20) cm × (0.4–1) cm, glabrous to pubescent; ligule ca. 1 mm. Inflorescence (3–10) × ca. 1 cm, burrs contiguous, rachis scabrous. Burrs globose, 0.4–1 cm, truncate, stipe pubescent, all spines and bristles retrorsely barbed; inner spines connate for 1/3–1/2 their length forming a globose cupule, the flattened free tips triangular, erect or inflexed, cupule and tips pubescent, outer spines in 2 divergent whorls, a median whorl of stout rigid spines equal to the inner teeth, and an outermost whorl of relatively few short, slender bristles. Spikelets 2–4 in burr, 4.5–7 mm; lower glume 1/2 spikelet length; upper glume 2/3–3/4 spikelet length.

Distribution: Pohnpei

Utilization: It is a gramineous forage plant with good palatability, and is suitable for grazing.

Plant

Inflorescence

Echinochloa colona

English: Awnless barnyard grass, Barnyard millet, Birds rice

Annual grass. Culms erect or ascending, up to 60 cm or more tall. Leaf sheaths compressed, keeled; leaf blades linear, flat, (3–20) cm × (0.3–0.7) cm, glabrous, sometimes with transverse purple bands, margins slightly scabrous, apex acute. Inflorescence narrow, 5–10 cm; racemes 1–2 cm, erect or sometimes stiffly diverging, simple, separated or overlapping by up to half their length or more, rachis usually without long, tubercle-based hairs, spikelets tightly congested in 4 neat rows. Spikelets plumply ovate-oblong, 2–3 mm, hirtellous, sharply acute; lower glume ca. 1/2 as long as spikelet; lower lemma staminate or sterile; upper lemma whitish at maturity, elliptic.

Distribution: Yapese, Pohnpei

Utilization: It is a gramineous forage plant with good palatability, and is suitable for grazing.

Inflorescence

Plant

Spikelet

Oplismenus compositus

Chuuk: Fatil

English: Armgrass, Basket grass, Running mountaingrass

Perennial grass. Culms stoloniferous, straggling, ascending to 20–80 cm. Leaf sheaths glabrous, pilose or tuberculate-hairy; leaf blades lanceolate to ovate-lanceolate, (3–20) cm × (0.5–3) cm, subglabrous to tuberculate-hairy, base usually oblique. Inflorescence axis 5–15 cm, glabrous to sparsely tuberculate-hairy; racemes 3–6, 2–6 cm, ascending to erect. Spikelets in 7–14 widely spaced, sometimes patent pairs, lanceolate, glabrous to thinly pilose; glumes herbaceous, subequal, awned; awns stout, green or purple, viscid; lower glume awn 5–10 mm; upper glume awn to 3 mm or occasionally absent; lower lemma subcoriaceous, 7–9-veined, acute or with a stout 0.3–1 mm mucro; upper lemma subcoriaceous, ca. 2.5 mm, smooth.

Distribution: Chuuk, Kosrae

Utilization: It is a graminous forage plant with good palatability, and is suitable for grazing.

Inflorescence

Plant

Spikelet

Digitaria microbachne

Annual grass. Culms tufted, decumbent, branching and rooting at lower nodes, 30–100 cm tall. Leaf sheaths glabrous or papillose-pilose; leaf blades linear-lanceolate, (5–20) cm × (0.3–1) cm, glabrous on both surfaces, papillose-pilose at base, base subrounded, apex acuminate; ligule membranous, 1–2 mm. Inflorescence digitate or subdigitate, axis 1–4 cm; racemes 5–12, stiff, 5–15 cm; spikelets paired, imbricate by about 2/3 their length; rachis triquetrous, narrowly winged, ca. 0.6 mm broad, margins scabrous. Spikelets narrowly lanceolate-oblong, 2–2.5(–3) mm, acute; lower glume absent or a minute rim; upper glume up to 1/3 as long as spikelet, 1–3-veined, margins ciliate, apex villous with overtopping hairs; lower lemma equal to spikelet, 5–7-veined, veins evenly spaced or with a wider interspace flanking the midvein, lateral intervein spaces and margins appressed pubescent to villous, rarely ciliate or setose; upper lemma yellowish to gray, subequaling lower lemma, acuminate.

Distribution: Pohnpei

Utilization: It is a gramineous forage plant with good palatability, and is suitable for grazing.

Inflorescence

Plant

Axonopus compressus

English: American carpet grass, Blanket grass, Broadleaf carpet grass

Perennial grass, with vigorous creeping stolons, forming sward. Culms 15–60 cm tall, compressed, nodes bearded. Leaf sheaths loose, strongly compressed, keeled, basal sheaths imbricate; leaf blades broadly linear to lanceolate, flat or folded, (5–20) cm × (0.6–1.2) cm, both surfaces glabrous or adaxial surface pilose, apex obtuse; ligule 0.3–0.5 mm. Racemes 2–5, digitate or subdigitate, 4–10 cm, only slightly diverging; rachis glabrous. Spikelets oblong-lanceolate, 2–2.7 mm, pilose or glabrous, apex acute; upper glume and lower lemma 2–4-veined, midvein absent, laterals marginal; upper lemma pale, oblong-elliptic, shorter than spikelet, obtuse with an apical tuft of hairs; stigmas pale.

Distribution: Yapese

Utilization: It is a dominant graminous species in natural grassland with higher turf value and it is suitable for grazing due to its high feeding value.

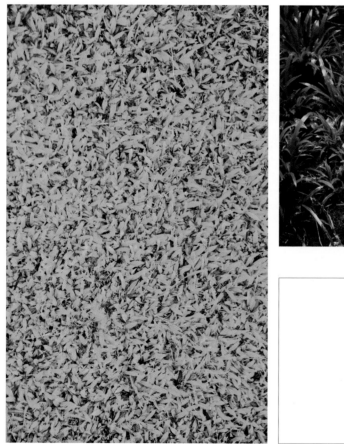

Lawn

Plantcluster

Spikelet

Paspalum scrobiculatum

Pohnpei: Ran-ta, Ranta

English: Creeping paspalum, Ditch millet, Indian paspalum

Perennial or annual. Culms tufted, slender to robust, erect or decumbent and rooting at lower nodes, 30–90 cm tall. Leaf sheaths compressed, keeled, usually glabrous; leaf blades linear or linear-lanceolate, (10–40) cm × (0.4–1.2) cm, usually glabrous, base subrounded, margins scabrous, apex acuminate; ligule 0.5–1 mm. Inflorescence of 2–5 racemes, subdigitate or on a short axis; racemes 3–10 cm, ascending to widely spreading; spikelets usually single, overlapping in 2 rows, sometimes paired especially in middle of raceme; rachis ribbonlike, 1.5–3 mm wide, margins scabrous. Spikelets green becoming brown, suborbicular, ovate or broadly elliptic, 2–3 mm, glabrous, obtuse to apiculate; upper glume membranous, 3–7-veined; lower lemma membranous or sometimes indurate, 3–5-veined; upper lemma brown at maturity, subequaling spikelet, coriaceous, finely striate, obtuse.

Distribution: Chuuk, Yapese, Pohnpei

Utilization: It is a gramineous forage plant with good palatability, and is suitable for cultivation as cut-and-carry forage or for rotational grazing.

Plant cluster

Plant

Paspalum paniculatum

English: Angel grass, Galmarra grass, Russell river grass

Perennial, forming coarse tussocks. Culms erect or geniculately ascending, 30–120 cm tall, nodes pubescent, sometimes rooting at lower nodes, longer than internodes. Leaf sheaths often hispid; leaf blades linear-lanceolate, flat, (9–50) cm × (0.6–2.5) cm, scabrid or hispid, margins usually undulate, apex acuminate; ligule 0.2–0.5 mm, membranous. Inflorescence axis 5–20 cm; racemes 7–60, fascicled, 4–12 cm, ascending or spreading; spikelets paired; rachis ca. 0.5 mm wide, winged. Spikelets brown at maturity, orbicular or obovate, 1–1.5 mm; upper glume membranous, subequaling spikelet, 3-veined, pubescent; lower gulme absent; lower lemma resembling upper glume, slightly pubescent; upper lemma as long as spikelet, pallid at maturity.

Distribution: Yapese, Chuuk, Kosrae, Pohnpei

Utilization: It is widely distributed in the Federated States of Micronesia, and often forms dominant populations in natural grassland. This grass yields high with a high leaf-to-stem ratio, and its leaves are tender, with high feeding value. It is one of the dominant graminous species for development of grassland and animal husbandry in this region.

Plant cluster

Inflorescence

Paspalum vaginatum

Pohnpei: Dimur, Dumwur, Timoor

English: Biscuit grass, Knot grass, Knottweed, Salt grass

Perennial with short rhizome and long stolons. Culms solitary or tufted, many-noded, 10–50 cm tall. Leaf sheaths imbricate, often keeled, about 3 cm long, margins membranous; leaf blades distichous, linear, rather stiffly ascending, (2.5–15) cm × (0.3–0.8) cm, apex acute; ligule 0.5–1 mm. Inflorescence of (1–)2(–3) racemes arising together at culm apex, opposite; racemes 2–5 cm, usually closely approximate when young, later spreading; spikelets single, in 2 rows; rachis 1–2 mm wide. Spikelets pale brownish green, narrowly lanceolate-oblong, strongly flattened, 3.5–4 mm, acute; lower glume absent or rarely a tiny vestige; upper glume thinly papery, weakly 5-veined, midvein often suppressed, glabrous; lower lemma resembling upper glume; upper lemma pale green, 2.5–3 mm, shorter than spikelet, cartilaginous, apex minutely white pubescent; anther about 1.2 mm.

Spikelet

Distribution: Yapese, Chuuk, Kosrae, Pohnpei

Utilization: It is a gramineous forage plant with good palatability, and is suitable for grazing.

Plant

Inflorescence

Paspalum dilatatum

English: Caterpillar grass, Dallis grass, Golden crown grass

Perennial from a short rhizome. Culms forming a coarse, spreading tuft, 50–150 cm tall, ca. 5 mm in diam., glabrous. Leaf sheaths glabrous or pilose in lower part; leaf blades linear, (10–45) cm × (0.3–1.2) cm, glabrous, apex attenuate; ligule 2–4 mm. Inflorescence axis 2–20 cm; racemes 2–10, 5–12 cm, spaced, diverging, axils pilose; spikelets paired; rachis 1–1.5 mm wide, glabrous. Spikelets green or purplish, broadly ovate, 3–4 mm, sharply acute; upper glume membranous, 5–9-veined, dorsally sparsely pubescent to almost glabrous, margins ciliate with long white hairs; lower lemma similar to upper glume, but eciliate; upper lemma pallid at maturity, orbicular, ca. 2 mm, clearly shorter than spikelet, papillose-striate, apex rounded.

Spikelet

Distribution: Pohnpei

Utilization: It is a gramineous forage plant with high yield and good palatability, and is suitable for cultivation as cut-and-carry forage, or for grassland improvement.

Inflorescence

Plant

Eriochloa procera

English: Slender cup grass, Spring grass, Tropical cupgrass

Short-lived perennial. Culms erect or geniculately ascending, branching, 30–150 cm tall, nodes pubescent. Leaf sheaths keeled, glabrous; leaf blades linear, (8–20) cm × (0.2–0.8) cm, glabrous, apex acuminate. Inflorescence 10–20 cm long, composed of racemes; racemes several, 3–7 cm, loosely ascending, bare of spikelets proximally; spikelets mostly paired, single toward raceme apex; axis and rachis very slender, puberulous, pedicels usually without setae, free or fused to each other. Spikelets lanceolate, 3–4 mm, herbaceous, sharply acute, basal swelling ca. 0.3 mm and often purplish; lower glume minute; upper glume and lower lemma 5-veined, pilose with appressed silky hairs, lower palea absent; upper lemma rugulose-punctulate, mucro 0.3–0.5 mm.

Spikelet

Distribution: Yapese, Chuuk, Kosrae, Pohnpei

Utilization: It is a gramineous forage plant with good palatability, and is suitable for grazing.

Plant

Inflorescence

Brachiaria mutica

Robust perennial. Culms stout, trailing, rooting freely from lower nodes, ascending to 2 m, 5–8 mm in diam., nodes, densely villous. Leaf sheaths villous or glabrous; leaf blades broadly linear, (10–30) cm × (1–2) cm, thinly pilose or subglabrous; ligule membranous, 1–1.3 mm. IPanicles 7–20 cm long, composed of 10–20 racemes; racemes 5–15 cm long, single, paired or grouped; rachis narrow, winged, scabrous; spikelets paired or single in upper part of raceme, in untidy rows or sometimes on short secondary branchlets in lower part of raceme; pedicels usually setose. Spikelets elliptic, green or purplish, 2.5–3.5 mm, glabrous, acute; lower glume triangular, 1/4–1/3 spikelet length, 1-veined; upper glume 5-veined; upper lemma rugulose, apex obtuse.

Distribution: Yapese

Utilization: It is a gramineous forage plant with high yield and good palatability, and is suitable for cultivation as cut-and-carry forage.

Plant

Node and Leaf sheaths

Spikelet

Panicum maximum

Chuuk: Paki ngeni

English: Buffalograss, Green panicgrass, Guinea grass

Perennial, rhizomatous; rhizome stout. Culms robust, erect, 1–3 m tall, nodes glabrous or pilose. Leaves basal and cauline; leaf sheaths glabrous to hispid; leaf blades linear to narrowly lanceolate, flat, (20–60) cm × (1–3.5) cm, narrowed at base, glabrous or pilose, margins scabrid, apex acuminate; ligule 1–3 mm, membranous, with dense cilia dorsally. Panicle oblong or pyramidal in outline, 10–45 cm, much branched; branches spreading, lowest arranged in a whorl. Spikelets oblong, 3–4.5 mm, glabrous or pubescent, green or tinged with purple, obtuse or acute, occasionally overtopped by long hairs from apex of pedicel; lower glume broadly ovate, 1/3–1/2 length of spikelet, 3-veined, obtuse or acute; upper glume ovate-oblong, as long as spikelet, 5-veined, acute; lower floret staminate, lemma similar to upper glume, palea well developed; upper floret thinly coriaceous, pale yellow or green, shiny, finely transverse rugulose.

Spikelet

Distribution: Yapese, Chuuk, Kosrae, Pohnpei

Utilization: It is a gramineous forage plant with high yield and good palatability, and is suitable for cultivation as cut-and-carry forage, or for grassland improvement.

Plant

Inflorescence

Panicum repens

English: Couch panicum, Creeping panic, Quack grass

Perennial, rhizomatous. Culms tough, erect or decumbent, 30–125 cm tall. Leaves cauline; leaf sheaths glabrous, striate, puberulous to ciliate on margins, especially toward throat; leaf blades linear, flat or convolute, often stiff and pungent, markedly distichous, ascending close to the culm, (7–25) cm × (0.2–0.8) cm, apex acute or acuminate; ligule 0.5–1.5 mm, a ciliolate membrane. Panicle terminal, narrowly oblong in outline, 5–20 cm, sparsely to moderately branched; branches glabrous, scabrid, ascending. Spikelets ovate, 2.5–3 mm, acute; lower glume broadly ovate, 1/3 length of spikelet, hyaline, 1(–3)-veined, clasping at the base of the spikelet, obtuse or acute; upper glume ovate, as long as spikelet, membranous, 7–9-veined, acute; lower floret staminate, lemma similar to upper glume, palea well developed; upper floret almost as long as spikelet, pale yellow, shiny.

Spikelet

Distribution: Yapese

Utilization: It is a graminous forage plant with good palatability and is suitable for grazing.

Inflorescence

Plant cluster

Coix lacryma-jobi

Chuuk: Fetin umuno

Pohnpei: Rosario

English: Adlay millet, Job's tears

Annual. Culms erect, robust, 1–3 m tall, more than 10-noded, branched. Leaves cauline; leaf sheaths shorter than internodes, glabrous; leaf blades linear-lanceolate, usually glabrous, (10–40) cm × (1.5–7) cm, midvein stout, base subrounded or cordate, margins scabrous, apex acute; ligule 0.6–1.2 mm. Male raceme 1.5–4 cm, spikelets in pairs with terminal triad; utricle ovoid to cylindrical, usually bony, shiny, (7–11) mm × (6–10) mm, white, bluish or gray-brown, sometimes with apical beak. Male spikelets oblong-ovate, 6–9 mm; glumes many-veined, lower glume winged on keels, wings 0.4–0.8 mm wide, wing margin ciliolate; anthers 4–5 mm.

Female spikelet

Distribution: Chuuk, Pohnpei

Utilization: It is a graminous forage plant with high yield and good quality, and is suitable for cultivation as cut-and-carry forage.

Inflorescence

Plant cluster

Miscanthus floridulus

Plant tufted, robust. Culms erect, 1.5–4 m tall, 6–15 mm in diam., unbranched, nodes usually glabrous, or uppermost sometimes bearded. Leaves cauline, congested; leaf sheaths longer than internodes, overlapping, glabrous, pilose at throat; leaf blades linear, flat, tough, (20–85) cm × (0.5–4) cm, glabrous, midrib prominent, margins scabrid, base rounded, apex acuminate; ligule 1–3 mm, dorsally densely pilose. Panicle oblong or elliptic, dense, 20–50 cm; axis 25–45 cm. Racemes numerous, 10–30 cm, appressed or ascending, glabrous, scaberulous; rachis internodes puberulous, nodes glabrous; lower pedicel 1–3.5 mm, upper pedicel 2.5–8 mm. Spikelets 2.5–4 mm, awned; callus hairs 4–6 mm, white, spreading, as long as the spikelet; glumes subequal, membranous, golden brown, 2.5–4 mm, margins pilose near apex, veins obscure, apex acuminate; lower lemma lanceolate, hyaline, 3–3.5 mm, veinless, pilose; upper lemma similar to lower, 2–2.5 mm; awn geniculate, 5–6 mm; upper palea a small hyaline scale. Anthers 3, 1–1.5 mm. Caryopsis oblong, ca. 1.5 mm.

Distribution: Pohnpei

Utilization: It yields high with general palatability, and is preferred by cattle and sheep when it is young and tender.

Plant

Inflorescence

Spikelet

Ischaemum polystachyum

Chuuk: Fatil

Pohnpei: Reh padil

English: Paddle grass

Perennial, rhizomatous. Culms loosely tufted, sometimes stoloniferous and rooting at lower nodes, 60–100 cm tall, nodes bearded or glabrous. Leaf sheaths glabrous or sparsely to densely pilose with tubercle-based hairs; leaf blades broadly linear, (5–20) cm × (0.5–1.5) cm, pubescent, rarely glabrescent, base rounded to subcordate, apex acute; ligule 1–2 mm. Racemes 3–6 or more, mostly terminal, subdigitate, 2–9 cm; rachis internodes and pedicels broadly linear, triquetrous, ciliate on outer angle, shortly ciliate on inner angles. Sessile spikelet lanceolate, (4–5) mm × (1.2–1.4) mm; lower glume leathery with expanded rounded flanks below middle, herbaceous, strongly veined and sharply 2-keeled above, glabrous or villous, keels usually winged, apex 2-toothed; upper glume attenuate into mucro or awnlet upto 2 mm; awn of upper lemma 1.2–1.5 cm. Pedicelled spikelet laterally compressed, similar to sessile, upper lemma awned.

Distribution: Yapese, Chuuk, Kosrae, Pohnpei

Utilization: It is distributed in four states of the Federated States of Micronesia, often grows on grassland as a dominant population. With high yield, tender leaves and extremely high feeding value, it can be cultivated as cut-and-carry forage, and also used to improve natural grassland for grazing. It is a dominant grass for the development of grassland and animal husbandry in this region.

Plant cluster

Inflorescence

Ischaemum ciliare

English: Batiki bluegrass, Indian murainagrass

Perennial. Culms slender, loosely tufted, erect, spreading or prostrate and rooting at lower nodes, up to 60 cm tall, nodes bearded. Leaf sheaths sparsely to densely pilose with tubercle-based hairs, or glabrous; leaf blades linear-lanceolate, (5–15) cm × (0.3–1) cm, tuberculate-villous or sometimes glabrous, base contracted, apex acuminate; ligule 1–2 mm. Racemes terminal, paired, often slightly separated, 2–9 cm; rachis internodes and pedicels oblong, triquetrous, ciliate along angles. Sessile spikelet obovate-oblong, (4–6) mm × (1.2–1.5) mm; lower glume smooth, glossy, leathery with rounded flanks in lower half, upper half flat, papyraceous, sometimes wrinkled, asperulous, flanks keeled, winged, wings 0.2–0.7 mm wide, forming 2 rounded lobes at apex; upper glume swollen and keeled above middle, keel narrowly winged, apex shortly awned; awn of upper lemma 1–1.5 cm. Pedicelled spikelet laterally compressed; lower glume with a single median winged keel; upper lemma awned.

Inflorescence

Distribution: Chuuk

Utilization: It usually forms a dominant population in mountain grassland, and is suitable for grazing.

Plant cluster

Bothriochloa ischaemum

Perennial, tussocky from a branching rootstock. Culms slender, erect or geniculately ascending, 25–70 cm tall, 3–6-noded, nodes glabrous or appressed bearded. Leaf sheaths keeled, congested at plant base; leaf blades linear, (5–16) cm × (0.2–0.3) cm, usually sparingly hairy with tubercle-based hairs, apex acuminate; ligule ca. 1 mm. Inflorescence composed of 5–15 racemes, subdigitate or inserted on a brief axis; racemes 3–7 cm, silvery-green or tinged with purplish brown; rachis internodes and pedicels ciliate with long white or pinkish silky hairs. Sessile spikelet 4–5 mm; lower glume oblong-lanceolate, usually cartilaginous, sometimes herbaceous, dorsally flat to slightly concave, 5–7-veined, silky-pilose below middle, lacking a pit, margins keeled and stiffly ciliate near apex; awn of upper lemma 1–1.5 cm. Pedicelled spikelet male or barren, subequal to sessile spikelet, glabrous.

Distribution: Pohnpei

Utilization: It is a graminous forage plant with good palatability, and is suitable for grazing.

Inflorescence

Plant cluster

Cymbopogon citratus

Perennial, shortly rhizomatous. Culms tufted, robust, up to 2 m tall, ca. 4 mm in diam., farinose below nodes. Leaf sheaths glabrous, greenish inside; leaf blades glaucous, (30–90) cm × (0.5–2) cm, both surfaces scabrid, base gradually narrowed, apex long acuminate; ligule ca. 1 mm. Spathate compound panicle large, lax, up to 50 cm, drooping, branches slender; spatheoles reddish or yellowish brown, 1.5–2 cm; racemes 1.5–2 cm; rachis internodes and pedicels 2.5–4 mm, loosely villous on margins; pedicel of homogamous pair not swollen. Sessile spikelet linear-lanceolate, (5–6) mm × ca. 0.7 mm; lower glume flat or slightly concave toward base, sharply 2-keeled, keels wingless, scabrid, veinless between keels; upper lemma narrow, entire and awnless, or slightly 2-lobed with ca. 0.2 mm mucro. Pedicelled spikelet 4–5 mm.

Distribution: Yapese, Chuuk, Kosrae, Pohnpei

Utilization: It is often used as an ornamental plant or spice.

Plant

Schoenus calostachyus

Perennials. Rhizomes short. Culms tufted, erect, 70–90 cm tall, obtusely 3-angled to subterete, sulcate, smooth. Leaves basal and cauline; sheath blackish purple; leaf blade linear, (30–75) cm × (1–2) mm, rigid, 3-veined on abaxial surface, margin scaberulose, apex acute. Involucral bracts leaflike; sheaths black purplish red to reddish black, 1.5–2.5 cm, terete; mouth pale, oblique, membranous. Inflorescences racemose, upto 50 cm, very lax, with 2 or 3 very distant fascicles of branches; peduncle 10–15 cm, glabrous; branches solitary or 2(or 3) together, erect, ca. 12.5 cm, unequal, compressed, scaberulose on angles, each bearing 1(–3) spikelets. Spikelets chestnut-brown, oblong to narrowly ovoid, 3–5-flowered, apex acute to acuminate. Glumes distichous, narrowly lanceolate, basal 5 and apical 1 or 2 empty; fertile glumes to 1.5 cm; sterile glumes ca. 0.4 cm, leathery, densely ciliate especially toward apex, keel green. Perianth bristles 4–7, white, ca. 1/2 as long as nutlet, antrorsely scabrous on apical part, easily caducous. Stamens 3; anthers linear, caducous. Style slender, apical half ciliate; stigmas 3, papillate. Nutlet grayish brown, obovoid, 3-sided, reticulately wrinkled, glabrous, base attenuate, apex obscurely beaked.

Distribution: Yapese

Utilization: It is a dominant species in coastal shrub grassland, and can be used for grazing.

Plant

Inflorescence

Rhynchospora rubra

Annuals or short-lived perennials. Rhizome short. Culms tufted, 30–65 cm tall, 0.8–2 mm in diam., terete, smooth. Leaves shorter than culm; sheath brownish straw–colored, 1–7 cm; leaf blade narrowly linear, 1.5–3.5 mm wide, papery, slightly scabrous, apex acuminate. Involucral bracts 4–10, stiffly spreading, leaflike, 1–5 cm, longer than inflorescence, unequal, densely ciliate at dilated base, sheathless. Inflorescence a single terminal head, brown to orangish brown, globose, 1–1.8 cm in diam., with many spikelets. Spikelets narrowly ovoid, 6–8 mm, shiny, 2–4-flowered. Glumes 7 or 8, brown, ovate-lanceolate to elliptic-ovate, thinly papery, keeled, vein 1, apex obtuse to acute; basal glumes each with a female flower; apical 1 or 2 glumes each with a male flower. Perianth bristles 4–6, unequal, 1/3–1/2 as long as nutlet, antrorsely scabrous. Stamens 2 or 3; filaments shorter to longer than subtending glume; anthers linear; connective evident. Style filiform; stigmas 2 or sometimes undivided, very short. Nutlet brown when mature, obovoid, 1.5–1.8 mm, biconvex, edges subacute with apical half hispid-serrulate, sides sparsely hispid-scabrous mainly on apical half, obscurely spotted with minute isodiametric epidermal cells; persistent style base conic, 1/5–1/4 as long as nutlet, base abruptly widened.

Plant

Distribution: Yapese

Utilization: It is a common companion graminous species in coastal shrub grassland, and can be used for grazing.

Inflorescence

Fimbristylis tristachya

Perennials. Rhizomes short. Culms densely tufted, 20–90 cm tall, 1–1.5 mm wide, flatly 3-angled, smooth, striate, with few leaves at base. Leaves shorter than culm; leaf blade 1.5–2 mm wide, stiff, involute, margin apically ciliate. Involucral bract 1, leaflike, much shorter than inflorescence, erect, margin ciliate. Inflorescence a simple anthela, with 3–6 spikelets. Spikelets ovoid to oblong-ovoid, (8–22) mm × (4–6) mm, terete, many flowered. Glumes spirally imbricate, brown with short rust-colored lines, ovate to broadly ovate, (5–6) mm × (4–4.5) mm, subleathery, with many veins, abaxial midvein slightly keeled, apex obtuse and mucronate. Stamens 3; anthers linear, 2–2.5 mm. Style ca. 3.5 mm, compressed, ciliate, basally slightly inflated; stigmas 2. Nutlet yellowish brown, stipitate, obovoid, ca. 2 mm, flatly biconvex, slightly shiny and with hexagonal reticulation.

Distribution: Yapese

Utilization: It can be used for grazing.

Plant

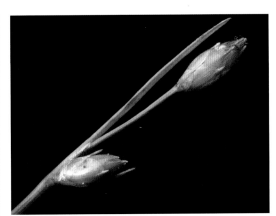

Spikelet

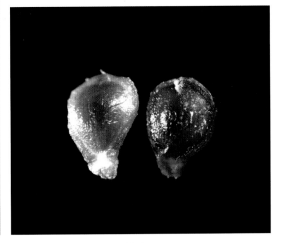

Nutlets

Fimbristylis cymosa

Perennials. Rhizomes short. Culms tufted, 10–60 cm tall, flatly and obtusely 3-angled, sometimes thick at base, with many leaves. Leaf blade 1–4 mm wide, thick, flat, extremely rigid, margin finely serrulate, apex acute. Involucral bracts 1–3, shorter than inflorescence. Inflorescence a simple or decompound anthela, headlike with a few short rays or open with several elongated rays. Spikelets numerous, solitary or clustered, oblong to ovoid, (3–6) mm × (1.5–2.5) mm, densely many flowered, apex obtuse. Glumes brown, ovate to broadly ovate, 1.2–2 mm, membranous, abaxially 3-veined,

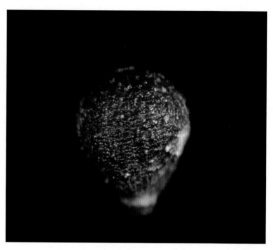

Nutlets

margin broadly hyaline, apex obtuse. Stamens 3; anthers linear. Style slender, not ciliate, basally slightly thickened; stigmas 2 or 3. Nutlet purplish black when mature, obovoid to broadly obovoid, 0.7–1 mm, 3-sided or biconvex, indistinctly verruculose and with square or transversely oblong reticulation but sometimes subsmooth.

Distribution: Yapese

Utilization: It can be used for grazing.

Population

Inflorescence

Bidens pilosa

English: Beggar's tick, Beggar-ticks, Black-jack

Annuals. Stems 20–180 cm tall. Petiole 10–30(–70) mm; leaf blade either ovate to lanceolate, (30–70) mm × (12–18) mm, or pinnately 1–lobed, primary lobes 3–7, ovate to lanceolate, (25–80) mm × (10–40) mm, both surfaces pilosulose to sparsely hirtellous or glabrate, bases truncate to cuneate, ultimate margin serrate or entire, usually ciliate, apices acute to attenuate. Synflorescence of solitary capitula or capitula in lax corymbs. Capitula radiate or discoid; peduncles 10–20 mm; calycular bracts 7–9, appressed, spatulate to linear, (3–)4–5 mm, abaxially usually hispidulous to puberulent, margins ciliate; involucres turbinate to campanulate, (5–6) mm × (6–8) mm; phyllaries 8 or 9, lanceolate to oblanceolate, 4–6 mm. Ray

Inflorescence

florets absent or 5–8; lamina whitish to pinkish, 5–15 mm. Disk florets 20–4; corollas yellowish, 3–5 mm. Outer achenes red–brown, ± flat, linear to narrowly cuneate, 4–5 mm, faces obscurely 2-grooved, sometimes tuberculate–hispidulous, margin antrorsely hispidulous, apex truncate or somewhat attenuate; inner achenes blackish, ± equally 4-angled, linear-fusiform, 7–16 mm, faces 2-grooved, tuberculate-hispidulous to sparsely strigillose, margin antrorsely hispidulous, apex attenuate; pappus absent, or of 2 or 3 erect to divergent, retrorsely barbed awns 2–4 mm.

Distribution: Pohnpei

Utilization: Pigs, cattle, sheep and poultry are all fond of feeding on it. It can be used as cut-and-carry forage.

Plant cluster

Crassocephalum crepidioides

English: Ebolo, Fireweed, Redflower ragleaf

Herbs erect, 20–120 cm tall. Stems striate, glabrous. Leaf petiole 2–2.5 cm; blade elliptic or oblong-elliptic, (7–12) cm × (4–5) cm, membranous, both surfaces glabrous or subglabrous, base cuneate, margin irregularly serrate or double-serrate, sometimes pinnately lobed at base, apex acuminate. Capitula several to numerous in terminal corymbiform cymes, shortly pedunculate, 3–5 mm in diam. Involucres cylindric, 1–1.2 cm, basally truncate, with few unequal linear bracteoles; bracts of calyculus 6–21, 2–6 mm; phyllaries uniseriate, linear–lanceolate, equal, ca. 1.5 mm wide, margin narrowly scarious, apically puberulent. Florets tubular, bisexual; corolla red-brownish or orange, rarely yellow, 8–10 mm; lobes 5, ca. 1 mm. Style branches acute, papillose. Achenes brownish, narrowly oblong, 1.8–2.3 mm, ribbed, hairy. Pappus 7–13 mm, early deciduous.

Distribution: Pohnpei

Utilization: Pigs, cattle, sheep and poultry are all fond of feeding on it. It can be used for planting as cut-and-carry forage.

Plant

Inflorescence

Tridax procumbens

English: Coat buttons, Tridax daisy, Wild daisy

Herbs, annual to perennial, caulescent, decumbent. Stems procumbent, branched at base, branches slender, spreading or ascending, 20–50 cm, hirsute. Leaves few, shortly petiolate; blade ovate to ovate-lanceolate, 2.5–5 cm, base cuneate, margin deeply irregularly serrate, pinnatisect, segments few, narrow, apex acute or acuminate. Capitula solitary, 1–1.5 cm in diam.; peduncle 10–30 cm; involucre subcampanulate, 6–7 mm; phyllaries few seriate, hispid, outer densely grayish white, elliptic, ca. 5 mm, long hirsute, inner tinged purplish, narrower, ca. 6 mm. Ray florets 4, white. Disk florets yellow, tube ca. 5 mm, limb 5-lobed, lobes reflexed, hairy. Achenes brown, oblong, ca. 2 mm, densely silky pubescent; pappus setae 5–6 mm, shiny, plumose.

Distribution: Pohnpei

Utilization: Pigs, cattle, sheep and poultry are all fond of feeding on it. It can be used for planting as cut-and-carry forage or grazing.

Plant

Inflorescence

Wollastonia biflora

Subshrubs herbs. Stems elongate, branched, scandent, coarsely appressed strigose. Cauline leaves long petiolate; petiole 1.2–2.3 cm; blade ovate, (7–14) cm × (3–8) cm, thickly papery, appressed strigose, base rounded, margin serrate, apex acuminate. Capitula (1 or) 3–6, terminal, 2–3 cm wide; peduncles 1.5–5.5(–8) cm, slender or thick; involucre (10–13) mm × (5–7) mm; phyllaries ovate-lanceolate or narrowly ovate, densely appressed strigose, gradually narrowed to tip. Ray florets 14 or 15, yellow, 1-seriate; corolla 9–13 mm, 2– or 3-dentate. Disk florets yellow; corolla ca. 5 mm, apex 5-lobed. Achenes (3–3.5) mm × (2–2.5) mm, cuneate at base, often 3-angled, coarsely strigose toward tip; pappus bristles 2 or 3, 2–2.5 mm, sometimes absent.

Distribution: Yapese, Chuuk, Kosrae, Pohnpei

Utilization: This species is highly invasive. However, it also has a feeding value. Pigs, cattle, sheep and poultry all prefer to feed on it. It can be used as a cut-and-carry forage due to its high biomass.

Inflorescence

Plant cluster

Elephantopus tomentosus

Herbs, perennial, 0.8–1 m tall. Rhizomes robust, obliquely ascending. Stems erect, multibranched, angled, white villous. Basal leaves withered at anthesis, subsessile or shortly petiolate, lower leaves oblong-obovate, (8–20) cm × (3–5) cm, abaxially densely villous and glandular, adaxially rugose and verrucose, sparsely or densely puberulent, basally progressively attenuate into winged, rather amplexicaul petiole, margin mucronate-serrate, rarely entire, apex acute; upper leaves elliptic or oblong-elliptic, (7–8) cm × (1.5–2) cm; uppermost very small. Synflorescence laxly corymbose, aggregated into compound

Plant

heads, surrounded by 3 leaflike bracts; bracts green, ovate-cordate. Capitula 12–20; peduncle slender, long. Involucre oblong, (8–10) mm × (1.5–2) mm; phyllaries green or sometimes apically purple-red, outer 4 lanceolate-oblong, 4–5 mm, glabrous or subglabrous, 1-veined, apex acute, inner 4 elliptic-oblong, 7–8 mm, sparsely appressed shortly hairy and glandular, 3-veined, apex acute. Florets 4, white, funnelform, 5–6 mm, with slender tube; lobes lanceolate, glabrous. Achenes oblong-linear, ca. 3 mm, 10-ribbed, puberulent. Pappus sordid white, of 5 basally widened bristles.

Distribution: Yapese

Utilization: Pigs, cattle, sheep and poultry are all fond of feeding on it. It can be used as cut-and-carry forage.

Inflorescence

Emilia sonchifolia

English: Purple sow thistle, Red tassel-flower

Herbs, annual; root vertical. Stems erect, gray-green, 25–40 cm tall, rather curved, usually branching from base, glabrous or sparsely pilose. Leaves thick, lower leaves crowded, abaxially dark green, often becoming purple, lyrate-pinnatilobed, (5–10) cm × (2.5–6.5) cm; terminal lobe large, broadly ovate-triangular, margin irregularly dentate, apex obtuse or subrounded; lateral lobes usually paired, oblong or oblong-lanceolate, both surfaces crisped-hairy, margin shallowly and bluntly dentate, apex obtuse or acute. Median stem leaves lax, sessile, smaller, ovate-lanceolate or oblong-lanceolate, basally hastately semiamplexicaul, margin entire or irregularly denticulate, apically acute; upper leaves few, linear. Capitula pendulous before anthesis, erect later, usually 2–5, in terminal lax corymbs; peduncles 2.5–5 cm, slender, not bracteate. Involucres cylindric, (6–12) mm × (1.5–4) mm; phyllaries 8 or 9, yellow-green, oblong-linear or linear, nearly equaling florets, glabrous, margin narrowly scarious, apically acuminate. Florets pink or purplish; corolla ca. 9 mm, with slender tube and dilated limb, deeply 5-lobed. Achenes cylindric, 3–4 mm, puberulent between ribs, 5-ribbed. Pappus of capillary-like bristles, snow white, ca. 8 mm.

Distribution: Pohnpei

Utilization: Pigs, cattle, sheep and poultry are all fond of feeding on it. It can be used as cut-and-carry forage.

Inflorescence

Plant

Ageratum conyzoides

Pohnpei: Pusen-koh, Pwisehnkou

Chuuk: Amshiip, Olloowaisiip, Ololopon

Herbs, annual, 50–100 cm tall, sometimes less than 10 cm, with inconspicuous main root. Stems robust, ca. 4 cm in diam. at base, simple or branched from middle, stems and branches reddish, or green toward apex, white powdery puberulent . Leaves often with axillary abortive buds; petiole 1–3 cm, densely white spreading villous; median leaves ovate, elliptic, or oblong, (3–8) cm × (2–5) cm; upper leaves gradually smaller, oblong, sometimes all leaves small, ca. 1 cm × 0.6 cm, both surfaces sparsely white puberulent and yellow gland-dotted, basally 3-veined or obscurely 5-veined, base obtuse or broadly cuneate, margin crenate-serrate, apex acute. Capitula small, 4-14, in dense terminal corymbs; peduncle 0.5–1.5 cm, powdery puberulent; involucre campanulate or hemispheric, ca. 5 mm in diam.; phyllaries 2-seriate, oblong or lanceolate-oblong, 3–4 mm, glabrous, margin lacerate; corollas 1.5–2.5 mm, glabrous or apically powdery puberulent; limb purplish, 5-lobed. Achenes black, 5-angled, 1.2–1.7 mm, sparsely white setiferous.

Inflorescence

Distribution: Yapese, Chuuk, Kosrae, Pohnpei

Utilization: Pigs, cattle, sheep and poultry are all fond of feeding on it. It can be used as cut-and-carry forage.

Plant cluster

Pilea microphylla

Pohnpei: Limw in tuhke, Limwin tuhke
English: Artillery fern, Artillery plant

Herbs weak, glabrous, monoecious. Stems erect or ascending, blue-green when dry, simple or branched, 3–17 cm tall, 1–1.5 mm in diam., succulent, cystoliths dense. Stipules persistent, triangular, ca. 0.5 mm, membranous; petiole slender, unequal in length, 1–4 mm; leaf blade abaxially pale green, adaxially green, obovate or spatulate, unequal in size, (3–7) mm × (1.5–3) mm, succulent, papery when dry, midvein indistinct distally, lateral veins several, indistinct, abaxial surface honeycombed, cystoliths linear, adaxial, regularly transverse, base cuneate or attenuate, margin entire, somewhat recurved, apex obtuse. Inflorescences often androgynous, compactly cymose-capitate; peduncle 1.5–6 mm, sometimes sessile; glomerules few flowered. Male flowers pedicellate, ca. 0.7 mm; perianth lobes ovate, subapically corniculate; rudimentary ovary minute, conic. Female perianth lobes subequal, oblong, longer lobe subequal to achene. Achene ovoid, ca. 0.4 mm, compressed, smooth, enclosed by persistent perianth.

Distribution: Yapese, Chuuk, Kosrae, Pohnpei

Utilization: Poultry is fond of feeding on it. It can be used as cut-and-carry forage.

Plant cluster

Pouzolzia zeylanica

Herbs perennial, erect, rarely prostrate, almost simple or few branched at base, 12–40 cm tall; rootstock often tuberous; branches sometimes with short branchlets, strigillose. Leaves often opposite, sometimes alternate on lower or upper stems; stipules triangular, 2–6 mm; petiole 0.2–1.8 cm; leaf blade ovate or broadly ovate, lanceolate or narrowly lanceolate, usually (1.2–9) cm × (0.8–3) cm, smallest ones on short branchlets, herbaceous, secondary vein 1 or 2 pairs, abaxial surface sparsely or sometimes densely strigillose or strigose along veins, adaxial surface glabrous or sparsely strigillose; base cuneate to rounded, rarely subcordate, margin entire, apex subobtuse, acuminate, or shortly so. Glomerules often bisexual, 2.5–5 mm in diam., bisexual ones in nodes of proximal leaves, female in distal axils; bracts triangular, 2–3 mm, ciliate. Male flowers: perianth lobes 4, narrowly oblong or oblong-oblanceolate, connate to middle, 1.2–1.5 mm, puberulent, apex acute or cuspidate. Female perianth tube ellipsoid or rhombic, 0.8–1 mm, 1.5–1.8 mm in fruit, puberulent, inconspicuously ca. 9-ribbed or 4-winged, apex 2-toothed. Achenes white, light to dark yellow or light brown, ovoid, 1–1.2 mm.

Distribution: Yapese

Utilization: Pigs, cattle, sheep and poultry are all fond of feeding on it. It can be used as cut-and-carry forage.

Plant

Ipomoea pes-caprae

Herbs perennial, glabrous, with a thick tap root. Stems 5–30 m, prostrate, sometimes twining, rooting at nodes. Petiole 2–10 cm; leaf blade ovate, elliptic, circular, reniform or ± quadrate to oblong, (3.5–9) cm × (3–10) cm, rather thick, 2-glandular abaxially, base broadly cuneate, truncate, or shallowly cordate, margin entire, apex emarginate or deeply 2-lobed, mucronulate. Inflorescences 1- to several flowered; peduncle stout, 4–14 cm; bracts early caducous, broadly deltate, 3–3.5 mm. Pedicel 2–2.5 cm. Sepals unequal, ± leathery, glabrous, apex obtuse, mucronulate; outer 2 ovate to elliptic, 5–8 mm, wider; inner 3 nearly circular and concave, 7–11 mm. Corolla purple or reddish purple, with a darker center, funnelform, 4–5 cm. Stamens included. Pistil included; ovary glabrous. Stigma 2-lobed. Capsule ± globular, 1.1–1.7 cm, glabrous, leathery. Seeds black, trigonous-globose, 7–8 mm, densely brownish tomentose.

Distribution: Yapese, Chuuk, Kosrae, Pohnpei

Utilization: Pigs, cattle, sheep and poultry are all fond of feeding on it. It can be used as cut-and-carry forage.

Population

Population

Vitex trifolia var. *simplicifolia*

Shrubs, 1.5–5 m tall, erect. Branchlets densely pubescent. Leaves 1–3-foliolate; petiole 1–3 cm; leaflets sessile, oblong, lanceolate, or obovate, abaxially densely gray tomentose, adaxially green and glabrous or subglabrous, base cuneate, margin entire, apex obtuse, veins ca. 8 pairs and slightly prominent on both surfaces; central or single leaflet (2.5–9) cm × (1.7–3) cm. Panicles 3–15 cm; peduncle densely gray tomentose. Calyx slightly 5-dentate, abaxially gray pubescent, adaxially glabrous. Corolla purplish to bluish purple, 6–10 mm, abaxially scaly white, pubescent at filament bases and on adaxial surface of lower lobe. Stamens exserted. Ovary glabrous, with or without glands. Style glabrous. Fruit black, subglobose, ca. 5 mm in diam.

Distribution: Yapese, Chuuk, Kosrae, Pohnpei

Utilization: Sheep like to feed on it and it can be used for grazing.

Plant

Inflorescence

Ipomoea aquatica

Chuuk: Aseri, Seeri, Seri
Pohnpei: Kangkong
Yapese: Kangking, Kangkong

Herbs annual, terrestrial and repent or floating; axial parts glabrous. Stems terete, thick, hollow, rooting at nodes. Petiole 3–14 cm, glabrous; leaf blade variable, ovate, ovate-lanceolate, oblong, or lanceolate, (3.5–17) cm × (0.9–8.5) cm, glabrous or rarely pilose, base cordate, sagittate or hastate, occasionally truncate, margin entire or undulate, apex acute or acuminate. Inflorescences 1–3-flowered; peduncle 1.5–9 cm, base pubescent; bracts squamiform, 1.5–2 mm. Pedicel 1.5–5 cm. Sepals subequal, glabrous; outer 2 ovate-oblong, 7–8 mm, margin whitish, thin, apex obtuse, mucronulate; inner 3 ovate-elliptic, ca. 8 mm. Corolla white, pink, or lilac, with a darker center, funnelform, 3.5–5 cm, glabrous. Stamens unequal, included. Pistil included; ovary conical, glabrous. Stigma 2-lobed. Capsule ovoid to globose, ca. 1 cm in diam. Seeds densely grayish pubescent, sometimes glabrous.

Distribution: Yapese, Chuuk, Kosrae, Pohnpei

Utilization: Pigs, cattle, sheep and poultry are all fond of feeding on it. It can be used as cut-and-carry forage, and is also an important vegetable resource.

Plant cluster

Inflorescence

Ipomoea batatas

Chuuk: Kómwu, Kómwuti, Kómwuutiy

Pohnpei: Pedehde, Satumaimo

Yapese: Gamuti, kamuut

Plant cluster

Herbs annual, with ellipsoid, fusiform or elongated subterranean tubers; sap milky; axial parts glabrous or pilose. Stems prostrate or ascending, rarely twining, green or purplish, much branched, rooting at nodes. Petiole 2.5–20 cm; leaf blade broadly ovate to circular, (4–13) cm × (3–13) cm, margin entire or palmately 3–5-lobed, herbaceous; lobes broadly ovate to linear-lanceolate, sparsely pilose or glabrous. Inflorescences 1– or 3–7-flowered; peduncle 2–10.5 cm, stout, angular; bracts early deciduous, lanceolate, 2–4 mm. Pedicel 2–10 mm. Sepals oblong or elliptic, ± unequal, glabrous or pilose abaxially, margin ciliate, apex acute, mucronulate, outer 2 sepals 7–10 mm, inner 3 sepals 8–11 mm. Corolla pink, white, pale purple to purple, with a darker center, campanulate to funnelform, 3–4 cm, glabrous. Stamens included. Pistil included; ovary pubescent or glabrous. Capsule rarely produced, ovoid or depressed globose. Seeds glabrous.

Distribution: Yapese, Chuuk, Kosrae, Pohnpei

Utilization: Pigs, cattle, sheep and poultry are all fond of feeding on it. It can be used as cut-and-carry forage.

Inflorescence

Asystasia gangetica

English: Asystasia, Chinese violet, Coromandel

Herbs to 0.5 m tall, ascending. Stems 4-angled, pilose. Petiole 3–5 mm, pubescent; leaf blade ovate to elliptic, (3–12) cm × (1–4) cm, glabrous or sparsely pilose especially on veins, adaxially with numerous cystoliths, base truncate to rounded, margin entire or slightly crenulate, apex acuminate. Racemes axillary or terminal, to 16 cm; bracts triangular, ca. 5 mm, pilose; bracteoles linear-lanceolate, 1–2.5 mm, pilose. Pedicel 2–3 mm, pilose. Calyx ca. 7 mm; lobes linear-lanceolate, (5–7) mm × (1–1.2) mm, abaxially glandular pilose, margin ciliate. Corolla yellow or white, 1.2–3.5 cm, abaxially gland-tipped pilose, adaxially glabrous; tube basally cylindric and ca. 3 mm wide for ca. 8 mm then gradually widened to ca. 1 cm; lobes obovate to semicircular, (0.7–1.2) cm × (0.8–1) cm; middle lobe of lower lip with violet or maroon markings. Stamens included; filaments glabrous, longer pair ca. 5 mm, shorter pair ca. 3 mm; anther thecae ca. 3 mm × 1 mm. Ovary ellipsoid, ca. 3.5 mm; style ca. 1.8 cm, velutinous; stigma slightly capitate, 2-lobed. Capsule ca. 1.3 cm × 2 cm, pubescent. Seeds irregularly obovate, (3–5) mm × (0.5–3) mm, tuberculate-rugose.

Distribution: Yapese, Chuuk, Kosrae, Pohnpei

Utilization: Pigs, cattle, sheep and poultry are all fond of feeding on it. It can be used as cut-and-carry forage.

Plant

Colocasia esculenta

English: Taro

Rhizome vertical to horizontal, tuberous, 3–5 cm in diam. Stolons long or absent. Leaves 2 or 3 or more; petiole green, 25–80 cm, sheathing for 1/3–2/3 length; leaf blade adaxially matte waxy-glaucous and water-shedding, oblong-ovate to suborbicular, (13–45) cm × (10–35) cm, base shallowly cordate, apex broadly and shortly cuspidate. Peduncle usually solitary, 16–26 cm. Spathe tube green, (3.5–5) cm × (1.2–1.5) cm; limb open proximally, cream-colored to golden yellow, lanceolate or elliptic, (10–19) cm × (2–5) cm, apex acuminate. Spadix: female zone conic, (3–3.5) cm × ca. 1.2 cm; ovary 1–3 mm in diam.; stigma subsessile, narrower than apex of ovary; sterile zone narrowly cylindric, 3–3.3 cm; sterile flowers (pistils) seen from above elongate, ca. 0.5 mm in diam.; male zone cylindric, (4–6.5) cm × ca. 7 mm; appendage narrowly conic, (15–45) mm × ca. 2 mm. Berry green, ca. 4 mm. Seeds few; synandria ca. 1 mm high, ca. 0.8 mm in diam.

Distribution: Yapese, Chuuk, Kosrae, Pohnpei

Utilization: Pigs and poultry are fond of feeding on its leaves and petioles.

Plant cluster

Cyrtosperma merkusii

Pohnpei: Mwahng; Mwating

Perennial large herb with tuber. Plant up to 6 meters tall. Leaves arising from short subterranean stems; petiole stout, up to 3 m long, erect, base usually softly spiniferous; large leaves ovate-sagittate, e, slightly oblique, up to 2.5 meters long, up to 1.5 meters wide, dark olive green, parted at base. Corms are large, and some species like Simihden produce corms up to 2 meters long and more than 25 kilograms/corm. Inflorescence solitary on a peduncle; spathe variable in size, color and shape; spadix 2–24 cm long, equaling or exceeding half the length of the spathe; Flowers hexamerous; ovary (1–) 2-ovulate; stamens exserted from the tepals at male anthesis. Fruit reddish orange, sessile, 1(–2)- seeded. It grows in freshwater swamps or in more humid areas and usually forms a high-density population.

Distribution: Yapese, Chuuk, Kosrae, Pohnpei

Utilization: Its corms are rich in starch and are a food crop with special cultural features in Micronesia. Leaf blade is huge and is an excellent source of feed for pigs and poultry.

Large leaf blades that can be used to keep off rain in the local area

Wild population

Spadix

Artocarpus communis

Trees 10–15 m tall, evergreen. Bark grayish brown, thick. Branchlets 0.5–1.5 cm thick. Stipules amplexicaul, lanceolate to broadly lanceolate, 10–25 cm, appressed pubescent with yellowish green, gray or brown hairs. Leaves spirally arranged; petiole 8–12 cm; leaf blade ovate to ovate-elliptic, 10–50 cm, thickly leathery, glabrous, abaxially pale green, adaxially dark green and shiny, margin entire, apex acuminate; secondary veins 10 on each side of midvein. Leaves on mature trees

Fruit

pinnately lobed or pinnatipartite; lobes or segments 3–8, lanceolate. Inflorescences axillary, solitary. Male inflorescences yellow, narrowly cylindric, narrowly ellipsoid, or clavate, 7–30 (–40) cm. Male flowers: calyx tubular, apically 2-lobed, pubescent, lobes lanceolate; anthers elliptic. Female flowers: calyx tubular; ovary ovoid; style long, apically 2-branched. Fruiting syncarp green to yellow, brown to black when mature, obovoid to ± globose, (15–30) cm × (8–15) cm, tuberculate; pericarp soft; mesocarp of milky white fleshy calyx. Stones ellipsoid to conic, ca. 2.5 cm in diam.

Distribution: Yapese, Chuuk, Kosrae, Pohnpei

Utilization: Pigs and poultry prefer to feed on its leaves not fruits.

Plant parts

Local use

Trema cannabina

Shrubs or small trees to 6 m tall, monoecious. Bark grayish brown, smooth. Branchlets green, brown, or purplish, variously pubescent or glabrescent. Stipules linear-lanceolate, 2–5 mm. Petiole 4–8 mm, slender, variously pubescent; leaf blade yellow-green to brownish green or brown (never blackish) when dry, ovate, ovate-oblong, or rarely lanceolate, (4–9) cm × (1.5–4) cm, base rounded to ± cordate or rarely broadly cuneate, margin crenate-serrate, apex acuminate to caudate-acuminate; basally 3-veined; secondary veins 2 or 3 on each side of midvein. Male inflorescences usually in proximal leaf axil of branchlets. Female inflorescences usually distal. Male flowers, pedicellate, ca. 1 mm in diam; tepals 5, obovate. Drupes subglobose or broad ovate, slightly compressed, reddish orange when mature, 2–3 mm in diam, with persistent perianth.

Distribution: Yapese

Utilization: Sheep like to feed on its leaves. It is suitable for grazing.

Plant part